Bamboo Rod Restoration Handbook

By Michael Sinclair

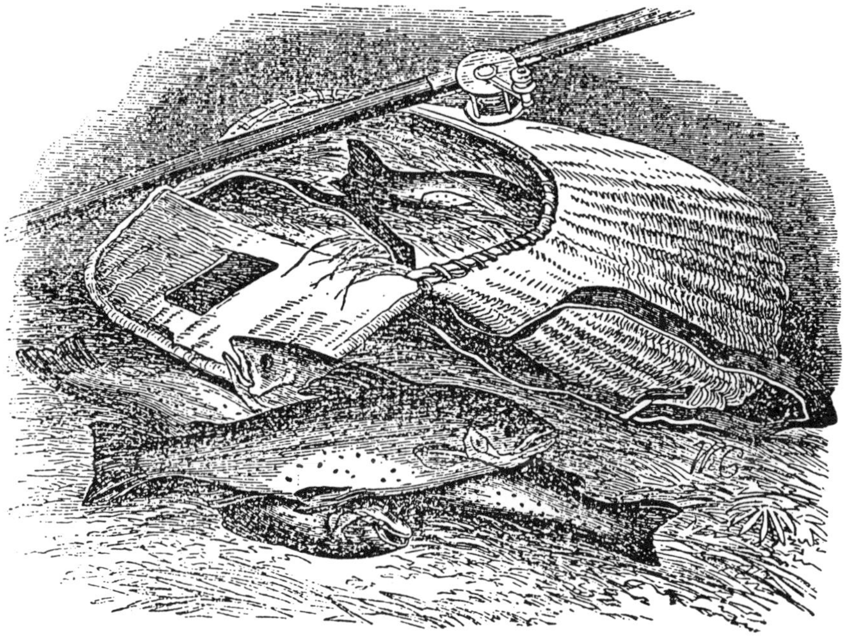

Complete With A Guide to Original Production Rod Specifications

Dedication

To rodmakers of the past and present, who live on through their creations; and to my little brother, Patrick, who has always been my fishing buddy.

Michael (left) and Patrick Sinclair
Young's Ranch – June, 1962 – Red River, New Mexico

Published by Centennial Publications
 ISBN 1–882418–11–5 – $25.00

 Address all inquiries to Centennial Publications, 256 Nashua Court - Grand Junction, Colorado 81503.

Contents

Cover Artwork: Reproduced from The Sportsman's Portfolio of American Field Sports
Published by M. M. Ballou of Boston in 1855

All Rod Illustrations by Michael Sinclair

Foreword – By Jon Parker

When I decided to make bamboo rods full time, one of the first things I did after selecting a business name and filing it with the county clerk, was to order business cards. In the center of these I placed my newly designed logo and in the left hand corner, just opposite my phone number, I had the word *RESTORATION* in bold block letters. Little did I know how much I would come to hate that word!

I was under the impression that if I could make a bamboo rod, I could easily handle the restoration of one. Subsequently I learned that restoration of bamboo rods requires more than technical *know how* for stripping, wrapping and varnishing.

I once read that many classic rods have been rendered worthless, both to their owners and to future generations, because the rods were refinished with no concern for originality. I did not want to join the ranks of restorers guilty of this oversight.

One of the first rods I received for restoration was a South Bend #359. The owner had started to do the job but soon tired of it. He made no notes prior to stripping the rod and, to make matters worse, had misplaced the guides he had removed. I could locate the guide spacing from the faded color of the cane but my only clue to the color of the guide wraps came from the owner, who remembered them as a peculiar shade of green. I began to realize there was a great deal I did not know about this rod. What type of stripper was originally on the rod? Was the tiptop wire or agate? Was it wrapped with silk or nylon? I really needed information.

I made some calls to fellow rod builders and, although meaning to be helpful, they replied to most of my questions with, *I think, I'm pretty sure, or It could be.* Everyone I contacted for information made educated guesses but none were certain. The only way to answer my questions with certainty was to find someone who owned an original #359 in mint condition. This was impossible. I was frustrated and did not know where to turn.

I once had a grade school teacher who, when questioned by a student about the location of a country or the spelling of a word, was fond of saying *knowledge is knowing where to look.* This would send us off to the atlas or dictionary to get the answer. However, unlike my fellow students, rod restorers found themselves at a roadblock. There was no place to look!

The questions of originality were not just a problem with the first rod I attempted to restore. There are unique problems and unanswered questions that cause frustration with every rod. My cowardly solution for avoiding this frustration was to order new cards and stationary without the "R" word in the left hand corner.

When Michael Sinclair first informed me of his plan to do this book, I was elated. It is a book desperately needed and long overdue. Whether it be the type of welt on a ferrule or the number of cork rings on a grip, this book will tell you in a glance. It will be beneficial to the professional as well as the amateur rod restorer. This book is also a valuable aid for identifying the make and model of unmarked rods.

It is no coincidence that the idea for this book came from Michael. In his restoration business, *The Cane Clinic,* he had encountered and solved numerous problems which helped him realize the need for a book dealing with restoration. The foundation of this book emanates from his extensive experience and his love of history and research. Michael is a dedicated rod historian, a persistent sleuth hunting for originality, and a top notch rod restorer. It is these qualifications that make this book so informative and helpful.

I am certain that every person who is a collector, restorer, or someone who just needs to know everything possible about his rod, will want to have a copy of this book. As my old teacher said, *Knowledge is knowing where to look,* and now it can be as close as your personal library shelf.

Jon W. Parker, Rodmaker
Ballston Lake, New York
August 18, 1994

Introduction

Years ago I bought an old, beat up, very used three-piece bamboo fly rod at a flea market. I had no idea where or when it was made, or who the maker might have been. I didn't know whether it was a good rod for fishing, or whether it would simply turn into a *wall-hanger*.

Since I've always enjoyed learning about new things I was looking forward to finding out more about the old rod. I went blissfully into the local library seeking enlightenment. I got it; but not in the way I had hoped. Instead I discovered there was no way to learn what I wanted to know from the books available. That kind of frustration can kill the budding interest of a novice. It didn't stop my interest from growing, but it did slow me down a lot.

One thing you find out quickly about cane rod *nuts* is that they're about the nicest and most helpful bunch of folks you'll ever meet. The trouble is that in trying to be helpful, sometimes they'll pass along faulty myths and inaccurate information they have heard from other equally well-meaning cane rod nuts.

I was lucky. A local fly shop directed me to Len Codella, who at that time was with Thomas & Thomas, and who has probably seen and handled more classic tackle than anyone I've ever met. I sent him a rough drawing of my rod along with a few notes to help describe it. Len was kind enough to let me know that I had a rod made by the Fred D. Divine Co. of Utica, New York. He *guessed* that the rod was made in the 1920's. His guess turned out to be right on the money; I later found out the rod was made in 1924. Len also told me that from my description, the rod did not sound like it was in original condition; that someone had probably refinished it. That was great, I thought. Now all I had to do was find out how the rod should look, and then find someone to put it back into its original condition for me.

In order to understand my position at the time you have to know that during that period in my life I was so broke, as they say, *I couldn't pay attention*. I was still sweating the $25 I had paid for my old rod, so you can imagine my surprise when I started trying to get it restored.

I literally *ran into* Steven Jenkins in one of the Denver fly shops. Call it fate; he was out trying to drum up restoration business. I showed him my old treasure. By this time, I was tired of carrying it around with the sections rubber-banded together, and had spent the last of my fortune for a cloth rod bag. Of course, the bag was for a nine-foot graphite rod and had only two sections and was two times too long for my rod. I still used the rubber bands, to keep the doubled over part under control.

I was dumber than dirt about cane rods, but Steven Jenkins took pity on me. He looked at my rod with genuine interest, which I have come to realize is normal for rodmakers. They always see more in a rod than I do, even today. Maybe they can feel the maker's fingerprints. They always get this far away look in their eyes, as if transported to a rod shop of a by-gone era.

After some careful consideration, Steven said that he was just beginning to get involved with the rod business, but he would talk to his father (Charlie Jenkins) about it and give me a call. When he did call back I was pleased and disappointed at the same time. It seemed they would be happy to do the work, but there was a problem. They had no idea what the original specifications were for a Divine rod.

Try checking out some old catalogs, they suggested, or look at some of the used rod lists to get some ideas for the wrap colors and patterns. As for the cost of restoration, they quoted me a price that was, in my condition, completely out of the question. I have since discovered that their price was actually quite reasonable and competitive. Steven said that the intermediate wraps the rod would require for originality would be an additional one dollar each. Have you ever seen a middle-era Divine rod? I counted the light marks on the cane and discovered that there were originally more than 100 intermediate wraps on my rod. The price of happiness was rising again.

The result of this long-winded tale is that I realized I would have to learn to restore rods if I wanted this one back as original.

There was a book on this subject at the public library. It is now out of print, but I highly recommend it for those who wish to learn to restore rods. That book is *The Fine Bamboo Rod: A Master's Secrets to Restoration* by Stuart Kirkfield. It is a great place to start, as it covers all the basic operations in a clear, concise and thorough manner. Let's put it this way; if Kirkfield could teach me, you'll certainly do fine with it. This book has some rare and valuable information, including an old *secret* formula for making an oxidizing agent for nickel silver hardware.

The information was there to tell me how to do restoration work. All that remained was putting it to practical use. Kirkfield was adamant about using silk thread for wrapping classic bamboo rods, even though nylon is much more widely available. I wrote him to ask where in the world I could find silk thread? *It may be quite common where you're from,* I told him, *but there's precious little of it in Denver.* Turns out that at that time nobody was making silk thread, and you had to hope for old silks still in stock in fabric stores and such. I had some luck, but not much.

My largest remaining problem was that nobody seemed to have any idea what the old Divine rod should look like. The wrap pattern was *ghosted* on the cane, so it wasn't hard to determine the pattern, but the colors of the wraps were absolutely unknown. I guessed. Red with gold tipping and gold intermediates.

I didn't use color preserver on the wraps. I put the hook keeper on the wrong flat. I didn't line up the witness marks when I mounted the guides. I over-varnished the cedar reel seat. It was the most beautiful cane rod I had ever seen. Of course, being the only cane rod I'd ever seen didn't hurt that evaluation. Months later, I received a photo of a Divine rod from a fellow wanting to sell it. The wrap pattern and colors were exactly as I had guessed. Go Figure.

If you have an old unmarked rod, like the one I bought, that you want to restore and you only know enough to know you don't know enough, this book is for you.

Rare and valuable classic rods by makers such as Payne, Gillum, Halstead, Hawes and the like deserve the best restoration talent available. They are certainly of high enough collectible and historic value that they are worth every dollar invested for a top notch restoration. Unfortunately, many other rods just don't have enough collectible or sentimental value to make an expensive restoration job feasible. There are many fine fishing rods that fall between the high and low ends that are destined for the stream rather than a showcase.

Examples include rods such as those made by Chubb, Montague, South Bend, Shakespeare, Horrocks-Ibbotson and most antique rods. On the border line are rods by makers such as Divine, Heddon, Goodwin Granger, Phillipson, and Wright & McGill.

Though information will be provided in this book on as many different makers as possible, the emphasis is on the vast majority of the rods – those that were, and still are, the ones most commonly encountered on the stream.

With this book you should be able to pin down the proper decade of manufacture for any unmarked rod. You can make a good, strong supposition as to its maker. And you won't have to guess how the rod originally looked. Its all in this book: distinguishing the maker, determining the original specs and recreating the original appearance of most rods. You'll find charts of wrap colors and patterns for most of each maker's various models. There is a chart of the proper colors of silk thread to match each maker's original specs.

Also included are sources for new silks, guides, ferrules and reel seats. Original specs will include the actual measurements and drawing of each maker's grip shapes and styles, original decals and handwritten rod markings as well as their proper locations on the rods. What you have in your hands is the accumulated knowledge that comes from handling, cataloging, studying, refurbishing, refinishing and restoring hundreds of rods over the years.

There are three fine books that deal with the specific techniques of making and restoring bamboo rods. There is no need for me to duplicate the information they provide in great detail. Instead, I will provide suggestions and instructions regarding how to apply the basic techniques those books contain specifically to the restoration of bamboo rods. First was *The Bamboo Rod and How to Build It* by Claude M. Kreider, originally published in 1951 and reprinted in 1992. More modern is *A Master's Guide to Building a Bamboo Fly Rod* by Everett Garrison with Hoagy B. Carmichael. The latest addition is the recently published book *Handcrafting Bamboo Fly Rods* by Wayne Cattanach.

In these books, the basic details of building a blank and finishing out a rod from a blank are covered in a thorough step-by-step manner. What has been lacking is a good step-by-step procedure guide for preparing an older rod for restoration. That's where this book steps in. There are two parts to restoring any old rod: *unbuilding* the rod down to its components, and then *re-building* the rod back to the original maker's specs.

With this handbook as your guide, you should find both a little easier than I did in the beginning.

Decisions, Decisions

There are many, many decisions to be made before you begin restoring a rod. The first, of course, is whether the rod should be restored *as original* or will instead be a refinished (or customized) rod. If the rod is to be as original, that's the only decision you need to make at this time.

If the rod will *not* be restored as original, you have decisions to make on every component you use and every rebuilding technique to be applied. To help you think through these decisions, here are some hints about items to consider. Is the rod for your own use, or are you doing the work for someone else? If you are working on another person's rod, let the rod owner make the decisions. There are many good reasons for this; self-defense being chief among them. Many customers will simply instruct you to *do what you think is best*. If that thought scares you, keep reading. When we are done you'll have a good idea what is best for the rod.

First, you must establish the eventual use of the rod. Is it for decoration? If this is its fate, the rod need not be anything more than cosmetically sound. This is a low-effort proposition. The ferrules need not fit together if the rod will never be assembled. It is easy to retain originality on these types of rod projects, and it makes a more interesting wall-hanger.

For fishing? The rod must be sound. The ferrules must be in great shape, even if they must be replaced; and the finish must be adequate to protect the rod. Cosmetic accuracy is optional. For collecting? This is the true test of your skill as a restorer. This rod must be restored *as original* to the greatest degree possible, and it should be mechanically sound for fishing as well. That's where the charts of original specifications can help.

Sentimental rods present an interesting situation. Most of these derive their value strictly from the previous owner. Many of them have no real dollar value otherwise. Some of them were customized by Grandads, Great Uncles and other angling *experts* to make them better tools. It is no easy decision whether to re-do the rod to the specs of the original maker, the previous owner, or to your own specs. I can't help you here, this is your decision.

Some folks ask me to leave the repair wraps on the rod just like Granpda left it. (Maybe they remember the day he was so excited about a nice catch that he stepped on his rod, or perhaps they remember him measuring his catch with the marks he made on the rod butt). Other interesting modifications are home-made hook keepers, custom carved grips and rulers that have been inked, painted or glued onto rod butts. I've even had some requests to leave the grip dirty. *It may just be sweat and fish slime*, the customer will say, *but it's Grandpa's sweat, and he caught the fish that provided the slime*. Many times, you'll have requests just to replace a wrap or a broken guide, and otherwise leave the rod as is. Customers are still always right, so let them tell you what they want.

If the rod is yours, *you* must make the decisions. To help decide the eventual purpose of the rod, ask yourself *What is it about this rod that appeals to me?* It might be the maker, the era of manufacture, sentimental appeal due to the previous owner, the appearance of the rod, or perhaps its value, either as a fishing tool or as a collectible.

Most likely it is a combination of more than one of these, but you must decide which stands above the others. Once you have decided what the most important factor is, set your sights on making the rod the best it can be for that purpose.

The largest number of decisions you'll face comes when the rod is yours, and you plan to keep it for fishing. You can strip such a rod down to glued-up sticks and make it entirely to your specs, or restore it as original. Most of the time, you'll find that you do some of both. Many rods that are otherwise great for fishing have very poor ferrules. If you use modern step-down ferrules you've hurt the collectibilty of the rod by a little, but you've made a giant step in making it a better fishing tool.

The biggest choices involve the pocketbook. Ferrules and the reelseat represent the largest portion of the cost of components, and have the widest range of prices. Just as 100 years ago, you will be faced with determining how much money you want to put into the rod. As you see the vast difference in the cost of the components, you'll realize why rodmakers charged such different amounts for the varying quality grades in their lines. Sometimes, the cost of new components makes you re-evaluate using the original parts; since you already own them the price is always right.

Just a note, never throw out any old rod components. No matter the condition, they are always useful in accurate restorations or for study. As time goes on, you'll build up an interesting *parts department*. You can often trade these to other restorers. It always seems that the other guy has just what you need and vice-versa.

Making many of these decisions in advance will save you a lot of time and frustration, and perhaps even some money. Make a plan, and work according to your plan. If you must order parts for the rod, try to order everything you'll need at one time. Go over your list of parts carefully to be sure you haven't forgotten anything.

When you are in doubt as to the best method or parts to use, try to keep it original, you can't go wrong that way.

Reasons for Restoration

Sometimes the simplest concepts are the most difficult to explain. There may be only one truth, but it seems to come in many sizes and colors. Originality and its value are the topic of many friendly discussions, and some not so friendly. Let's start with the Webster's version of what original means: "coming from someone as the originator, maker, author, etc." Taking this definition one step further, the term *as original* should be interpreted to mean "in the style of, or as from the originator or maker." To keep it simple, I live with an easy rule *If it isn't all original, it isn't original at all.*

Original condition does not mean that the rod looks as good as the day it left the factory. It does mean that it still has all of its original parts and has not been customized nor had newer components added. All of the hardware and guides, the grip and reelseat, the wraps and the finish must be the originals. A new coat of varnish skillfully applied does not significantly harm the originality of the rod if everything else is original. Any other changes make it *customized* or *refinished*.

Because there are so many facets of collecting and using classic tackle, people with different interests have widely divergent views on the value of originality. For the pure collector, originality is absolutely priceless. The same is true for those who, like me, have an interest in studying the methods and styles of the classic rodmakers of the past. It is difficult, if not impossible to learn much from a *non-original* rod. On the other hand, if intended for fishing duty, the most pressing concern is for the rod to be sound and in good fishing condition. Originality isn't much of an issue in this case. Making the rod usable for fishing may preclude use of the original guides which might be very worn, impossible to find, or too small to accommodate today's fly lines.

Original condition is vital for determining the methods of construction used by the rodmakers. Through the methods used, rod historians can establish a recognizable style. This is similar to a police detective establishing an *M.O.* for a suspect, which may eventually lead to identifying the actual culprit.

The three most common changes to a rod's original condition are the reelseat, guides and silk wrappings. These were usually modernized at the time of a refinish. The location, pattern and color of the silk wraps can help solidify an identification of the origin of an unmarked rod. Unless a rod was refinished at the factory, it is unlikely that the original style of the maker was preserved. Even if it was done by the original maker it is likely that some updated hardware was added. There is also a good chance that the wrap color and pattern might be changed.

While the rod may arguably still be in the style of the original maker, it will not be true to its original specifications. Such a rod could send mixed signals as to its era of manufacture, and would be quite perplexing.

Many rods were marked on the rod shaft with ink or fragile decals, and these markings are often lost in careless refinishing. Classic all nickel silver sliding band style reelseats usually carried the maker's name engraved on the metal, and the name of the maker might not have appeared anywhere else on the rod. As new styles of reelseats and lighter, more modern materials became popular, many of these older style reelseats were replaced, and the maker's name was therefore removed from the rod. Almost every rod manufacturer made his reelseats available through his catalog or dealers, and it is so easy to remove the old nickel silver reelseats that many amateurs did the work themselves. While this may have made them better fishing tools, it did damage the value for study and reference.

The question of whether to preserve the originality of the rod may depend on its intended use. For the record, I believe that the best rule is always to keep every rod as original as possible.

I don't care who you are, or what you think you know – the original maker did things the way he did for a reason, and since he was the boss in his shop, he was accustomed to getting his way. I believe we owe it to the classic rodmakers to honor their original plans. Call it simple respect. Even if the reason a rodmaker did something his way just to save a buck, he had a reason. Live with it. If as a refinisher or *handyman* you decide to *improve* the rod with simpler and easier wraps, different silk colors (or worse, using nylon thread), by using polyurethane instead of spar varnish, or taking any other modern shortcut, just imagine that old rodmaker standing over your shoulder asking, *just what the hell do you think you're doing?*

The degree of originality that is preserved in re-doing a rod determines whether the rod is *restored* or *refinished*. If a rod is truly restored, you should be unable to distinguish the new work from a mint-condition rod from the original maker. Any change from the original specs is a refinish. It is very important to know what the original specs were for a rod to ensure an accurate restoration.

With the proper knowledge, a restorer can begin with a rod in non-original condition and return it to *as original* specs. To do so, the restorer must be knowledgeable about the methods used by the maker; the manufacturer and style of hardware and guides; number of guides and the proper sizes; the exact color, patterns and locations of all wraps; the precise markings on the rod shaft, the marking methods and locations of the markings; the methods used to seal and preserve the wraps; and the type of finish the maker used. If the cork grip is to be replaced, the restorer needs to know the style of the grip, width of the cork rings used by the maker, length of the grip and the dimensions for turning down the grip.

If the restorer doesn't know these specs, he must locate a rod of the same era by the proper maker and of the proper model to study for the clues it can provide. Let's hope he finds one that truly is in *original* condition. Remember, though, that in truth any restored rod is not *original*, but *as original*.

Due to an invincible enemy – time – many, many rod components from the classic era are collector's items themselves. It can be impossible to locate some guide styles, many of the original spec ferrules, and even more so, replacement bamboo sections to replace missing or broken rod joints. Some replacement tip sections are available, but on a limited basis. This presents a challenge to the restorer. Some rodmakers will accept the challenge of reproducing rod joints by hand planing a replacement. These are no bargain to buy, but are unavailable through almost any other source. Some ferrules can be duplicated by talented metalsmiths or machinists, but again, the cost is very high. A standard style step-down ferrule is commercially available, but is not in the exact style of any particular maker. Thus it is excusable for some rods to be just a little short of an actual restoration. If the parts needed to preserve originality just are not available, it seems best to come as close as possible to the original. For example, Perfection tip-tops were patented in 1915. At that time, the patent date was roll-stamped on every tip-top along the N/S tube, "Pat 8-3-15". Later, the company logo was used as a stamping. After about 1955, no marking was placed on the tip-top. The latest change came in the late 1970's with the addition of a *large loop* tip-top to the company's line. Until the large loop came into being, there was no real discernible difference in the appearance and style of the Perfection tip-top. This means that, though not *exactly* the same, any Perfection tip-top would be acceptable, as long as the small loop was used. To some degree, the same small compromises may be made with ferrules and guides without straying too far from the original design, though it will never be exactly *as original*.

Attempt to reproduce to the greatest degree possible the methods used by the old rodmakers. The most simple part of maintaining originality is in recreating the appearance of the genuine wraps, both in appearance and in execution. Wraps were usually *color preserved* by the classic rodmakers. In order to maintain originality of appearance, the same method used by the maker must be used by the restorer to recreate the original cosmetics of the rod. Take a trip back in time. Companies that produce some of the modern miracle products of today such as Gudebrod and U-40 did not exist at the turn of the twentieth century. That simply means you can't use the modern color preservers. The old rodmakers certainly didn't know what a *water-based polymer color preserver* was. Don't misunderstand, this is a fine product for sealing wraps; it just isn't right for classic restorations.

For that matter, nylon thread is much easier to find and easier to use than silk thread, but it just isn't appropriate for an honest attempt at restoration. If you want to recreate the original, accurate and genuine appearance of a classic rod, you must use the same materials and supplies used at the time the rod was made. For color preserver, use clear lacquer or shellac, whichever the original maker used. These are still readily available and of little cost, so don't make excuses; just do it right. Remember, it isn't restoration if it isn't *as original*. It's just a refinish. How do you know which was used? You must study an original condition rod from that maker and from the proper era and use trial and error to recreate the appearance of the original condition rod. Now you see why it's important to rod historians and restorers to have some original rods extant!

Here's another good point for studying what is original. Have you ever seen a South Bend *Comficient Grip,* the Wes Jordan Rod grip from Orvis, or the grip on a Marvin Hedge Tournament Rod made by Heddon? If you've never seen one, you would probably think that these grips could not possibly be original. That's because they have some unique features such as carved-out thumb rests, palm rests and such that give the rod the look of a poorly sheared sheep. But these are original. And lots of amateurs, in the spirit of do-it-yourself rod improvement, dutifully carved their own thumb rests and such. Some of these are factory, some are butchery. You must study originals to know which is which.

The concept of originality can be carried to an extreme; for some, if a rod has had a polish on the varnish, it is no longer original. For others, as long as the rod is recognizable as to maker, it is original. The truth probably lies somewhere in the middle. Since it is difficult to find agreement in terms of what is *original,* and what originality is worth, I'll use Ben Franklin's quote. He was asked which religion was the *true religion,* and Franklin replied, "Different sects are like different clocks; They may all be near the matter, though they don't quite agree."

Rods Worth Restoring

All bamboo rods are a part of our angling heritage. As such, they all have some value, even if it is only sentimental. For many rods, it is their value in a historic sense that makes them valuable. They may not be suitable for fishing in today's styles. Indeed, they may deserve a place in a museum. Consider a collection of antique or historically significant rods as your own little museum and you can't go wrong.

The decision of whether or not to restore an antique rod can be a difficult one. The originality of such rods should be left unimpeached in every possible way. Perform only work that is necessary to preserve the integrity of the rod for future generations. In many ways, the charm of these rods *is* their original condition. Time does some lovely things to cane rods that simply cannot be duplicated. The patina on German silver, the faded silks, the tea-stained or ambered varnish and the antique fittings all contribute an ambience of antiquity that is unmatched by anything we can create today.

Under normal circumstances, little or nothing should be done to an antique rod. Sometimes it is necessary to find a way to preserve old, cracked and dried out silks just to prevent them from deteriorating further. Usually, that is the only work that should be done on antiques. I cannot recommend any other procedures as they would be performed at the expense of the historic significance of the rod. If you are in possession of an antique that is in such poor condition that it must be restored, this book will offer some suggestions for trying to maintain as much of the original appearance as possible and still protect the rod.

For some rods, the value is strictly sentimental. It doesn't matter that the rod's current retail value is negligible; if it was Grandad's rod, it is valuable to you. Many fishermen did streamside repairs to rods, or customized them to better suit their needs. These changes can add some charm and interest to an otherwise run-of-the-mill rod. Sometimes it is best to leave these as is, especially for sentimental reasons.

For collectible rods, originality is most desirable. If a rod isn't *as original* you will probably want to return it to its original specs. This will make it more valuable, both collectibly and historically.

Many production-quality cane rods are used strictly as fishing tools. Rods destined to be *fishing rods* do not need to be all original to have value. Making a classic era rod fishable may preclude use of some classic older style guides which tend to run small in diameter since silk lines were of smaller diameters than modern fly lines. Additionally, guides are usually worn to some degree and may require replacement for fishing. These are the rods that you may refinish to your own tastes without substantially harming their value.

If you copy the style of a *better* maker in refinishing a rod, it is imperative that you never try to mislead a potential buyer about the true history of the rod. Always keep accurate notes on your work so you may give a proper accounting of any replaced parts, rod sections, or new cosmetics. As always, honesty is the best policy.

All rods are worth restoring. The value may be perceived in ways other than how many dollars it would fetch at sale, but the value is there nonetheless. The tougher decision is whether restoration is the best course for the rod. Consider the points made above, and decide what is best for the rod's intended purpose. One point to consider is that you can always decide to restore a rod later, but once the work is begun, you can never recreate its current condition.

Identifying and Dating Unmarked Rods

You've just acquired a bamboo fly rod, and all you can be sure of is that it is made of bamboo. There are no markings on the reelseat hardware, nor on the rod shaft. Naturally, this rod didn't come in a tube with a handy label. So who made it?

That's the first thing everyone wants to know, and if you want to restore a rod correctly it is essential to find out. The maker is a prime determining factor of the collectible and historic value of a rod. I wish I could tell you that there is a good set of rules to follow, but there isn't. There are some *rules of thumb* that can help, especially in dating a rod; but for every rule there is an exception. That's okay though, it keeps things interesting.

Very few rods ever left the factory with no markings of any kind. Those that did were usually of the lowest quality and are of little collectible value. They are interesting, and a part of rod making and fishing history, but they have little value otherwise. When we speak of unmarked rods, we include those that have a model name, but do not carry the maker's name. Rods like these are called *Trade Rods* because they were made *for the trade*. They may very well carry the name of the seller, and a model name.

Although companies such as Abbey & Imbrie, Von Lengerke & Detmold, Von Lengerke & Antoine, New York Sporting Goods, Dave Cook, Sears and Roebuck, Montgomery Wards and countless others sold thousands of rods bearing their names, none of them ever actually produced rods.

Even the fine old William Mills & Son Standard Rods were not of their own manufacture. These were actually made for Wm. Mills by the H. L. Leonard Rod Co. That means that Leonard was a maker of trade rods. Although Abercrombie & Fitch sold rods made by many makers, their top of the line rods were made by E. F. Payne Rod Co. These were marked with the name of the maker, and thus are not considered to be trade rods.

Some rodmakers worked through Sole Agents. The rods sold in this manner are not considered trade rods, because they carried the maker's name on the rod. An example of these were the rods sold by F. E. Thomas through *Frank E. Phillips, Sole Agent, Boston* or the Fred D. Divine Co. rods sold by *A. G. Spaulding & Bros, Western Agents, Denver*.

Trade rods were made by most large rod companies, from Horrocks-Ibbotson to H. L. Leonard. Sporting goods dealers needed many different quality levels to appeal to more fishermen's wallets. Trade rods sold from less than one dollar to nearly $100. Thus most dealers bought rods from more than one maker to fill different price niches.

Identifying the makers of some trade rods is pretty simple, while others can be very puzzling. Other than the markings on the rod shaft, many trade rods were identical to the normal models sold by the rodmaker. Sometimes the wrap colors or patterns were changed, or a different reelseat was used. This was not an attempt to disguise the maker, but was instead an effort to individualize the rod to meet the needs of the trade buyer, especially his cost goals and the eventual selling price. In almost every case, any change from the maker's regular specs was a down grade to reduce the price. Some trade rods were *seconds* in quality, and just didn't meet the stringent quality controls of the maker.

Among the makers whose trade rods are easiest to identify are Wright & McGill, Edwards, F. E. Thomas, H. L. Leonard, Heddon and South Bend. These rods are easy to spot because they retain more than 90 percent of the characteristics of marked rods by these makers.

The trade rods that are hardest to identify are usually those of the lowest quality, and are therefore the most numerous. These include rods made by Chubb, Montague, Horrocks-Ibbotson and Divine.

Still, once you know the general characteristics of the various makers, it isn't that difficult to make an educated guess.

To begin identification, it is always a good idea to try to figure out the period of time when the rod was built. This is actually fairly easy when dealing with high quality makers, but can be much more difficult in the case of the less well-made rods. The folks who made the trade rods often used old parts left over by quality rod makers. The better makers usually adopted new ideas in hardware and materials more quickly, and their inventories of less modern parts were sold to producers of trade rods. Additionally, many companies sold their hardware for use by amateur rod builders, and this complicates trying to determine the actual maker.

Just to clear up a point, there are many ways to decide which rods were so-called *Production Rods*. In this book, the term *production rods* refers to those that were made by mass production methods in very large quantities. That means all trade rods fall into the production rod category.

To begin identifying and dating a rod you must assume the rod you are evaluating is all original. You may later decide that it is not, but you've got to start somewhere. Over the years, many rods have been *restored, or refinished* to the point that the original maker wouldn't know his own work. The most common areas to be changed were the guides, wrap colors, wrap patterns and ferrules. Face it, rods do wear out to a certain extent, and from time to time they require a *re-do*. The least likely part of any rod to be changed is the grip and reel seat. Let's begin there.

Grips and Reelseats

Older rods have longer reelseats and longer buttcaps. That's a fact, and one of the few things you can count on in your evaluation. Older reelseats tend to be all metal, either nickel silver or nickel plated brass. They usually have a sliding ring to secure the reel to the reelseat. Sometimes you'll get lucky and there will be a patent date on the reelseat hardware. That's a help, but not necessarily the end of your search for the maker. Many patents were assigned (sold) by the patentee to someone else. Also, the patent date merely tells you the *oldest* the rod can possibly be. If it has an 1880 patent date, it certainly doesn't mean that's when the rod was made; it just means that it couldn't have been made any earlier. There are many patent dates from the late 1800's and early 1900's that were still marked on rods 50 years later. That's because 50 years is the usual period granted to the patentee for exclusive rights.

Measure the length of the reelseat. If it is over 4" long, it's a safe bet that it is Pre-1930. If it measures over 4½" long it is probably between 1900 and 1920. Any reelseat that is longer that 5" is likely to have been made between 1875 and 1899. Now for the exception; some older reelseats were not necessarily long, but these are few. So the rule of thumb is, *Longer is Older*.

Next examine the grip. What material was used? Wood grips were common until 1882, when Charles F. Orvis introduced cork covered grips. The wood grips were usually of sumac, sometimes smooth and sometimes ribbed or beaded. The ribs may run lengthwise or may be arranged in a pattern that appear like rings around the grip. Some were painted black, others were left natural and varnished. Another common grip on early rods was wood wrapped with rattan. This is usually a sign of a well-made rod in the early days, but not so much so after 1900. The standard rattan wrapped grips had red thread wound in the grooves of the rattan wrappings, creating an attractive spiral pattern in the grip.

If the grip is covered with cork, look at it carefully. Orvis used six strips of sheet cork, each about 3/8" wide, running lengthwise on the grip. Later, sheet cork was usually wrapped in a single piece around a wood base. If this is the case, you should find only one seam in the cork.

Cork rings come in many variations. The earliest cork rings were only 1/8" to 1/4" in width. That was the most readily available thickness for the sheets of cork from which the rings were cut. If you are evaluating a rod with narrow cork rings such as these, you can bet that you have a rod made from 1890 to 1930. Later, after the early 1900's, the standard width for most cork rings became 3/8" to 1/2" for better rods, and remains so today. Many unmarked rods will have wider rings, some up to one full inch in width. These are certain to be production rods, and not particularly well-made.

Another bit of information is in order: the two ways cork is cut into rings. Refer to the drawing below for the difference in these methods for cutting rings from cork sheets. The best cork is

referred to as *split jar specie cork*. It is cut "with the grain" so that the gaps run from butt to tip on the rod. In the drawing, the examples on the left demonstrate split jar specie cork rings. Other methods cut the cork rings across the grain, allowing a direct path for water to pass through the grip to the cane beneath. Remember that since bamboo is really a grass (even though it is generally referred to as a wood), it is bio-degradable. Thus, keeping moisture out was of paramount importance.

The better rod makers were extremely careful in selecting the cork for the grips, the best used split jar specie cork, and selected the rings with the fewest *lacunae*, or gaps that could allow moisture into the grip. Also it makes a more attractive grip. Usually, the lower grade rods received the rejects, those not good enough for the top-of-the-line models. It almost goes without saying that split jar specie cork was more costly to obtain.

By examining the grip and reel seat, you should be able to roughly determine the period when the rod was made. This will eliminate some makers, being either too early or too late for their production. You can also begin to get a feel for the overall quality of the rod. Then, additional information can be gained from studying the remaining hardware: the winding check, the guides, the ferrules and the tiptops.

Winding Checks, Ferrules & Guides

Winding checks were usually larger and more ornate on early rods. Some were simple but elegant nickel silver rings located where the grip meets the bamboo. Others were ornately incised. Some are very useful in identifying the maker, as they are virtually a signature piece. The important thing to remember is to note that although there are many different types of winding checks, the style used by most makers tended to remain fairly constant.

Ferrules, too, are valuable in helping determine the maker. Nickel silver is the material of choice for the highest quality ferrules. This is because it is virtually non-corrosive, and is a *slick* metal, ideal for marking parts requiring close tolerances for tight fits. The shape of the ferrule can be a clue to the maker, but the usual reference is the *welt* at the mouth of the female ferrule. This welt may be a ring, or rings, soldered to the ferrule, or it may be a *rolled* edge. It is not difficult to tell which was done. The better ferrules are those with welts that were soldered on, or actually machined as an integral part of the ferrule.

The ferrules can be the best clue to the maker of any rod. The problem with ferrules is that they did wear and needed to be replaced. Some cheaper ferrules also had a tendency to crack, and these were routinely replaced, and not always with the original type or style. In evaluating any rod it is important to determine if the ferrules are original or replacements.

Some ferrules were inscribed with patent dates; notably those of H. L. Leonard and those used on the Kosmic rods. These are patent dates from the 1870's and 1880's. About the only other marking usually seen on ferrules is the logo for the Super Z ferrule patented in 1952.

Next look at the guides, which are a good determiner of the age of the rod, if they are original. If there are different types on the same rod, it's a good guess that the older style was the original and the new style guides are replacements. The oldest of the guide styles is the loose ring type. Some people refer to these as *flip-ring or floppy-ring* guides. They are very similar in style to the classic hook keeper, a simple ring held in place by a strap of nickel silver which is secured by the guide wraps. This style was in common use until the very early 1900's.

The next guides to gain popularity were the familiar snake guides. However, the first snake guides to be used were produced on the English pattern. These guides are twisted in reverse to those produced in America, and many people refer to them as *reverse twist guides*. English style guides were in common use in this country until about 1925. By that time, most of the better makers had switched to the American style guide. One notable exception was the F. E. Thomas Rod Co. of Bangor, Maine. Thomas used the reverse-twist guides throughout the history of the company, into the 1950's. F. E. Thomas was certainly one of the better rod makers, so here is the exception to the guide rule.

The wire used for making guides is another indicator of age and quality. The older rods that exhibit snake guides usually used nickel silver wire. The problem was that this metal was a bit too soft and the fly line formed grooves in the guides. Better makers switched to bronze wire and the nickel silver guides were relegated to the lesser makers by about 1925. In the 1920's, the Perfection Tip Co. of Denver, Colorado, began producing guides made of chrome plated steel, and these were quickly recognized by the better makers as being the best guide.

In general, the best way to determine the age of snake guides is by the size. Smaller dimensions were used on older rods, as the silk lines in common use were of smaller diameter than the synthetic lines that came into use in the 50's. Many rods from 1900 into the 1940's had snake guides in sizes down to 4/0 on the tips. By comparison, today it is more common to find that the smallest snake guides on a rod are a size #1.

Prior to the time when stripping guides were made of chrome plated steel and other very hard metals, most stripping guides for fly rods and bait rod guides were nickel silver wire frames

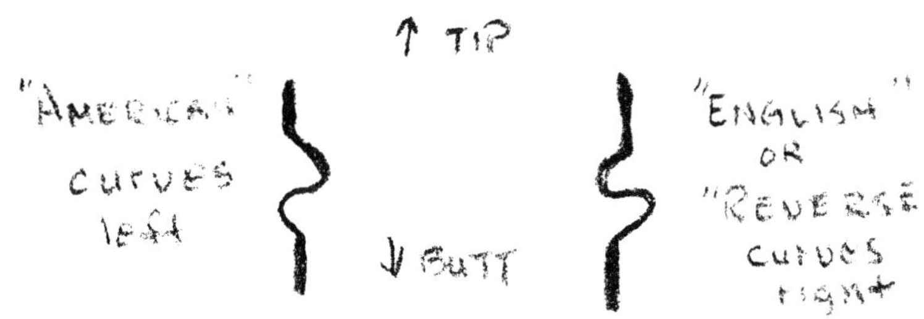

holding a ring of polished agate. These were not only functional but were also beautiful. The tiptops were also made with agate rings. These guides may be very old, but they are not always a reliable factor in dating rods. In the 1930's, guides resembling agate guide were produced from a variety of materials, including glass. These artificial guides were marketed as *Real Garnix* by Granger, and as *Agatine* by South Bend, Montague and Horrocks-Ibbotson. These synthetic lining materials tend to look *too perfect* to be agate, and are not as durable.

Tip tops are somewhat useful for dating rods. The earliest all metal tiptops had round loops. Perfection Tip Co. introduced tiptops with small pear-shaped loops in 1915. Larger pear-shaped loops were introduced in the 1960's.

Bamboo

The final indicator of the general age of a rod is the type of cane used to make it. Although there are many varieties of bamboo, only two types have been important to rod makers. The earliest of the two types used in constructing bamboo rods is called *Calcutta Cane*. This type is easily recognized. The natural color is very light tan, sometimes as pale as ivory, with randomly dispersed dark patches scattered throughout its culms. The dark patches are actually char-marks. Calcutta cane was often crooked and worm-infested, so after harvesting, the surface of the stalks of cane was burned to kill the worms. While warm, the culms of cane were straightened. The dark patches are marks left by this burning.

About 1900, *Tonkin Cane* became popular. This cane came from China, not from the Gulf of Tonkin area in southeast Asia. Tonkin Cane was a trade name used by Horrocks-Ibbotson Rod Co. for Tsing-Li cane. The name Tonkin soon became the generic term for rod making bamboo, much the same as Coke became the generic name for all cola.

Tonkin cane had many advantages over Calcutta cane, among these is the almost total absence of worms, and culms that grew impressively straight. This eliminated the need for burning the cane, and the result was a clear, straw colored cane with few imperfections. The lack of imperfections made more of the cane usable in constructing high quality rods, resulting in less waste. Tonkin cane also had excellent tensile properties, making it ideal for the flexing demands placed on fishing rods. By 1910, most of the makers of quality rods had switched to Tonkin cane.

By placing your rod into the period in which it was manufactured, you are well on the way to determining who actually made the rod. I should stress that although the maker and era are the two most important factors in determining the collectible value of a rod, neither of these has anything to do with how good the rod is for fishing. Remember that just as our grandfathers were more likely to own a Ford than a Cadillac, they were more likely to own a Montague than a Payne. While many people still swear by the Horrocks-Ibbotson rods, others would only swear at the same rod. It's all a matter of point of view. Years ago, just as today, you could spend $5 for a rod, or $500 or more. Either way, you could catch fish and enjoy fly fishing.

Determining the Maker

One of the easiest ways to begin determining the maker of a rod is to try to place an approximate date of manufacture on the rod. If original, ferrules are a good determiner of the age of a rod. In many cases they may also *tell* us who made the rod.

Simply put, ferrules are cylindrical metal joints that connect the sections of a rod. Methods for connecting two sections of a rod have undergone many changes through the years, and the steps along the way are good indicators of the period in which the rod was made.

The ferrules that were in wide use from about 1800 to about 1890 were built on a principle called *doweling*. These ferrules are very easily identified by the tapered dowel (also called pin or spike) that protrudes from the bottom of the male ferrule. This dowel fits into a recess in the female ferrule, and provides a longer (and thus it was hoped, a smoother) transference of the stress at the joint of the rod.

The earliest of these ferrules were created simply by *rolling* a piece of metal around the bamboo or wooden dowel and ferrule station, and being soldered at the seam. This construction caused the wood at the base of the ferrule and the dowel to be exposed.

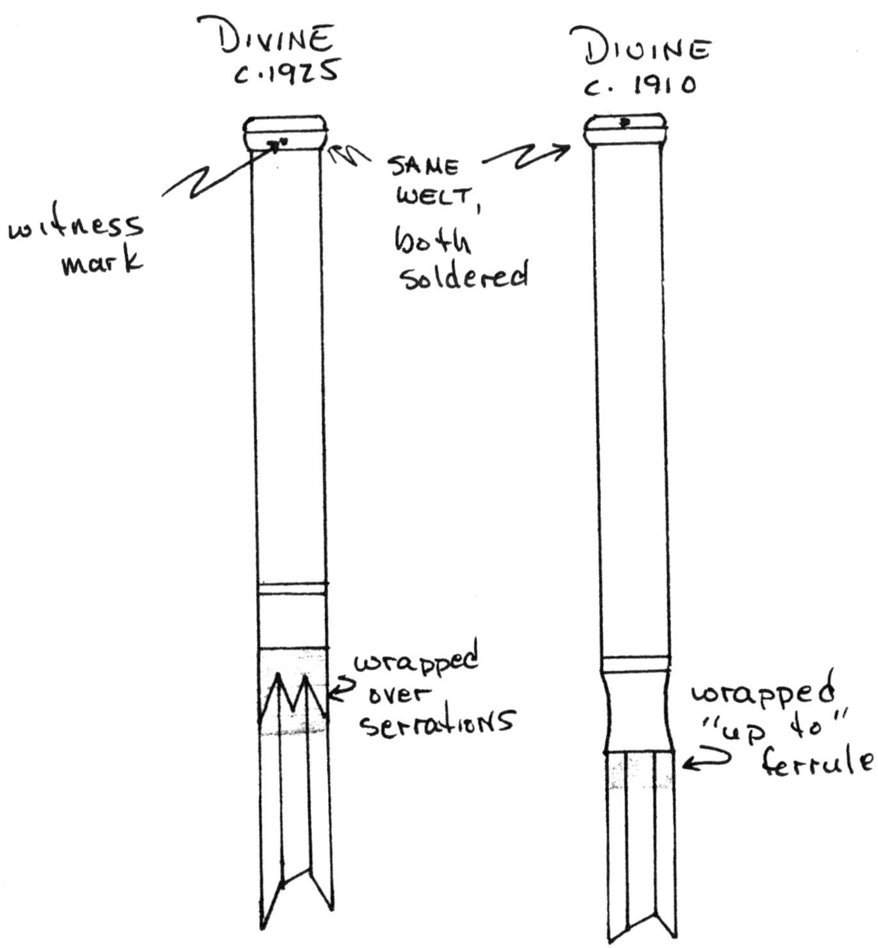

The next step in the evolution of the ferrule was the addition of a ring at the mouth of the female ferrule called a *welt*. These reinforced the opening and made it more resistant to being bent out-of-round. This addition began to appear in around 1860.

The first *modern* approach to ferrule design came about just after 1860 and involved the usage of two pieces of tubing of different diameters – the smaller fit inside the larger tube, and the tubes were then soldered at this joint. This created what was called a *step* or *shoulder*.

At this time, it is necessary to make a few observations about the material used and the quality of the ferrules. The best ferrules have always been those created from nickel silver (also called *German silver*, or from World War I, *Liberty silver*). Nickel silver does not actually contain any silver, but is more accurately called *white brass*, since it is brass (an alloy of zinc and copper) with nickel added to make up 18% of the total compound, which gives it a silvery appearance. Nickel silver is virtually oxidation-proof with minor care, and is an extremely *slick* metal, ideal for producing non-stick ferrules. Additionally this material also provides the strength that is needed while still providing an acceptable amount of flex in the joint.

While cheap fittings on production rods are nearly useless in determining age, I have noted that the production companies used the same type of ferrules for long periods of time, and that they are useful in determining the maker of a rod.

Later developments in ferrule construction began with Hiram Leonard's patents of 1875 and 1878. These patents solved two major problems. The earlier patent sealed the bottom of the male ferrule and the bottom of the dowel to make them waterproof. At the same time, a *cup* was soldered inside the female ferrule to make it waterproof as well. The later patent involved creating a series of thin cuts in the edge of the ferrule where it met the rodshaft, This made the ferrule flex with the rod, rather than making the stiff edge a fulcrum point that could cause the rodshaft to break.

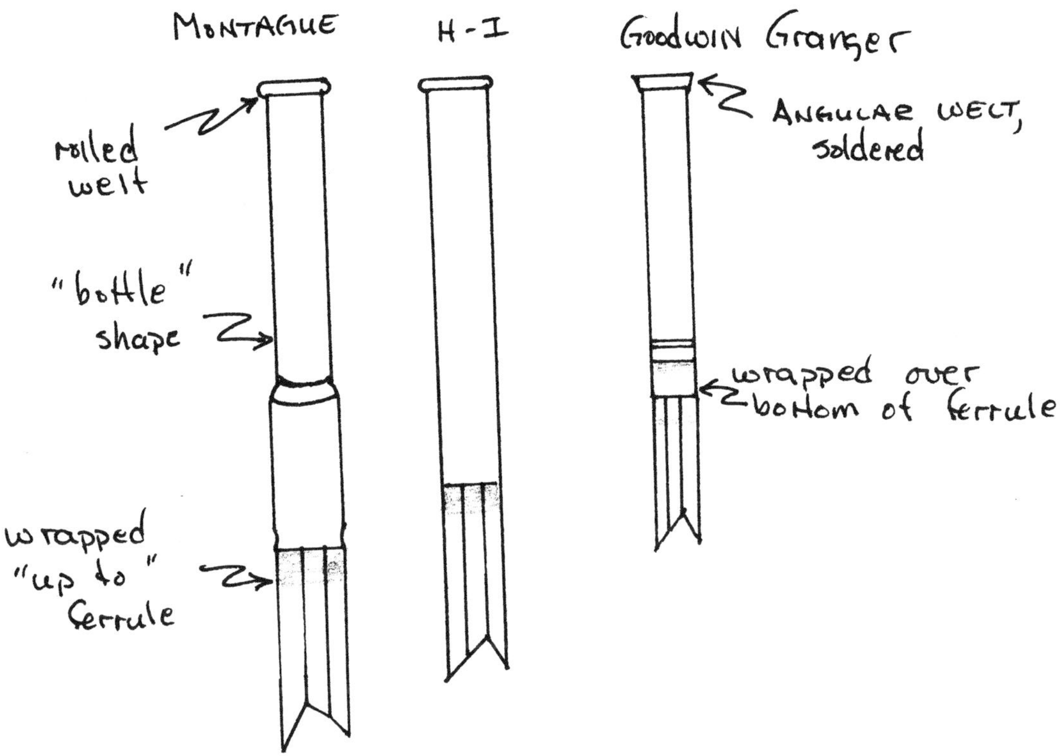

By 1889, most of the high quality ferrules were both waterproofed and serrated in some fashion to take advantage of Leonard's improvements, and most ferrules were made without the dowel. This type of ferrule is the most commonly seen, and is referred to as a *simple* ferrule.

The next major change in ferrules came after about 1880 when Thomas Chubb of Post Mills, Vermont, began to create his reelseats by drawing metal into a seamless tube with a cup-shaped base. This process was quickly adapted for making ferrules and made the standardization of ferrule sizes possible. Female ferrules were made waterproof simply by soldering a disc inside the tube. The ease of this method of construction made ferrule production much simpler and faster. It is easy to determine whether the ferrule was made from tubing (machined), or was *drawn*. The machined ferrules, since it was made from tubing, had a flat bottom. A disc of nickel silver was soldered into the end of the tubing to make it waterproof. The drawn ferrule began as a sheet of nickel silver, and the process of drawing the sheet into a ferrule produced a slightly *cupped* bottom. See the drawing for comparison of these types; the difference is readily identified, and helps date the ferrule.

With all of these improvements in place, the evolution of the ferrule as we know it was complete. While there have been many variations and experiments, the ferrule itself has remained largely unchanged, in both design and material.

Unmarked rods bearing their original ferrules have a good chance of being identified as to maker because of the many cosmetic possibilities to embellish the appearance of ferrules. Although there were variations within some maker's lines (i.e., using a lower quality ferrule on lower grades of rods) most are characteristic of a particular maker. This is true even with production rods made by the larger rod companies.

The first clue as to maker is whether the ferrule is of the machined or drawn type of construction. Production rodmakers after 1900 almost exclusively used drawn ferrules because they could be produced in large quantities in standard sizes.

Male ferrules that are of the drawn type have very few differences to distin-

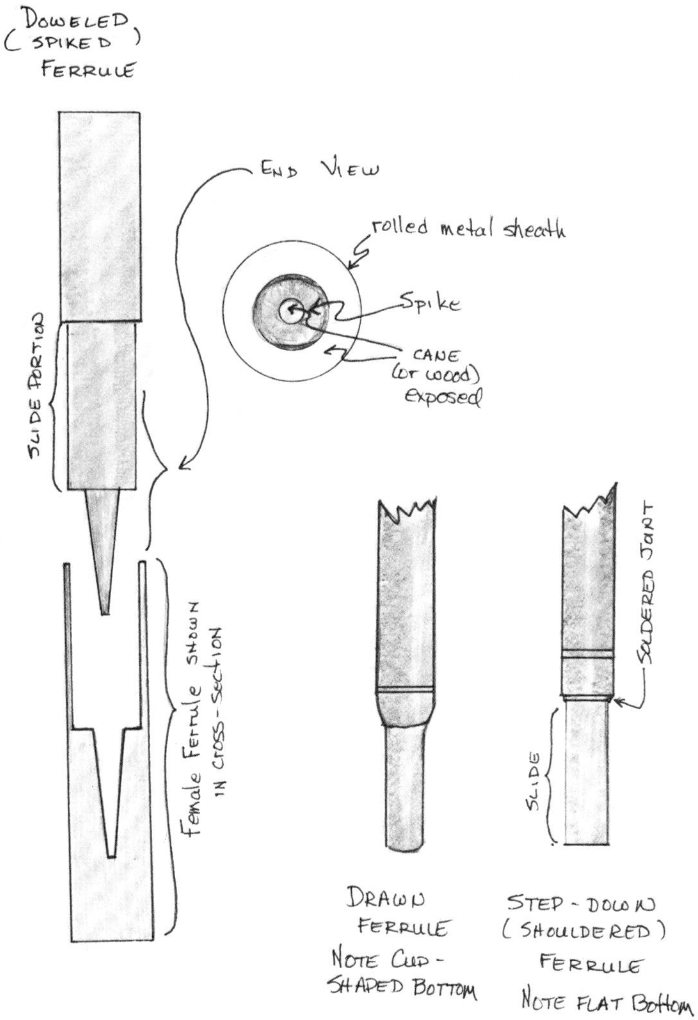

guish them. Usually, if they are marked at all it is with simple incised bands, which serve no function other than decoration. The only other distinguishing feature is whether the ferrule is designed to have the thread wrapped up over the edge of the ferrule where it meets the bamboo, or whether it is designed to have the thread come just up to the bottom of the ferrule.

Thus it is primarily the shape of the female ferrule that makes it possible to help determine the maker. There are two general shapes that are usually seen, either the *straight* design, or the *bottle-shaped* style. Most of the cheaper rods manufactured by Montague, for example, have had the bottle-shaped ferrules. Most of the Horrocks-Ibbotson rods had very plain, straight ferrules, some with an incised line near the rod shaft, but most with no mark at all.

More clues to the maker can be gained from the type of welt the female has around the upper opening, if one is present at all. Welts were of two general types. They were either *rolled* edges of the ferrule, or they were individual rings that were soldered onto the ferrule. Rolled welts indicate rods of lesser quality. Soldered welts indicate a bit higher quality ferrule, that was more time consuming and expensive to construct, and were generally found on ferrules used on higher quality rods. Ferrules that are *machined* and have a soldered welt are often referred to as *hand-made ferrules*.

Very few changes have been made in ferrules over the past fifty or so years, with one major exception. That exception is the Super Z ferrule patented in 1952 by Louis Feierabend. He is an engineer who used his skill and knowledge to create what may be the *perfect* design for a bamboo rod ferrule. The reason for this is that the ferrule's design allows the bamboo to remain at its full dimension at the ferrule, the weakest part of any rod. Older designs demanded that the bamboo must be *shaved* off at this critical location, thus creating a weak point where the rod was certain to suffer the greatest demands of stress. Furthermore, the Super Z ferrule allowed a precise continuation of the taper of the rod, resulting in a smoother transfer of action. Many of the premier rodmakers, notably Everett Garrison, immediately saw these advantages and used these ferrules exclusively.

As for the American evolution of the ferrule, that pretty well ends the discussion. However the ferrules produced in Great Britain are another story.

British Rod Features

The British seemed unable to convince themselves that the friction ferrule produced in the United States was reliable. (The British called them *suction ferrules*).

Most British ferrules were made of bronze or brass rather than nickel silver. Early advertisements for the oft Rube Goldberg British ferrules stated that these complex designs would relieve the angler from the worry of *flinging off a rod joint during casting*. Although I don't believe that many of us worry excessively about that, the ads assure me that this is *a constant source of anxiety for the dry-fly caster*.

Among the designs produced in Great Britain are the Locking Collar and the Hardy Patent Twist Lock ferrules. The former involved sliding a pin on the male into a slot on the female, then twisting the pin into a notch. The whole mechanism was then secured by the collar that was slid into a position that would hold the pin in the slot. The twist-lock Hardy ferrule had a *fin* with a notch that extended from the male ferrule. On the female ferrule there was a *thread* that was engaged by the slot, and the male ferrule was then *screwed* onto the female. A later Hardy patent ferrule had the same fin and slot, but the fin was spring loaded so that it would actually clip over an extra collar just below the ferrule's welt.

All of these ferrules are like a banana, easy to show but difficult to describe. However, if you encounter any of these ferrules on an unmarked rod it is a dead give away that the rod is of British origin. All of this fiddling around could have been eliminated by simply making a close tolerance friction-fit ferrule on the American design. It is possible, however, that the British simply got caught up in the idea that permeated the fishing tackle businesses of that era – everything had to be *new or improved*. Competition was extremely tough from 1880 to 1930, and perhaps these complicated designs were just another marketing effort to stay ahead of the competition.

British rods, whether by Hardy, Farlow, Sharpes, Allcock, Milwards, or any other of the many English makers, tend to be much alike. Hardy Brothers was the most prolific and perhaps the best known to American anglers. Here are a few of the characteristics they shared. They usually have only one tip (a real angler needed only one it was claimed). The guides were generally what we call the reverse-twist or English style, and usually made of heavy bronze wire. The rods usually featured hundreds of intermediate wraps very closely spaced; in fact, many people still refer to intermediate wraps as English winds. The action of the typical British rods was usually rather stiff with heavy tips; some refer to them as *clubby*.

To meet the desire of British anglers for even stiffer rods, Foster Brothers produced a rod that featured *steel ribbed cladding*, which was steel wire wrapped the entire length of the rod in such a way as to appear in a criss-cross pattern. This was in response to rods produced by several companies that had a Steele-Centre. I suppose that the ultimate stiff British rod would have been the rods made by Hardy that had a steel center and were also double-built! I have often wondered if the stiff rod was an extension of the *stiff upper lip*, or perhaps a Freudian explanation would be more accurate. It probably doesn't matter except to the British.

The rods and reels produced for many many years by Hardy are considered to be the absolute standard for high quality British fishing tackle. Even if Americans did not seem to care for many of the rods, we have always thought highly of the reels such as the Perfect and St. George models.

Grips

Quite a few people have asked me "What does a *coke-bottle grip* look like?" It should be obvious, I thought. But after two or three dozen inquiries on grip shapes, I began to realize that there's a lot of confusion. I've often been asked if a rod can be identified (or if at least some makers could be eliminated) by the grip shape. The answer is a definite maybe. So far, for all of the rods I have seen or read about, it seems that just about every maker has used just about every grip shape at some point in his career. There are exceptions, of course, but not enough to mention.

I have a habit of *beginning at the beginning* so here goes. The earliest grips seem to have evolved simply as fatter sections near the butt end of the rod. Of course these earliest grips (or *handles* as they were then called) were simply the largest diameter of the rod material, usually wood. There was nothing fancy about them. After having fished a circa 1860 14-foot wood rod (that weighed *pounds,* not ounces) I can imagine some genius thinking that it might be wise to do *something* at the place where you *grip* the rod to make it a little easier to control.

Two methods seemed to become popular. First was to turn the *handle* section of the butt on a lathe to make grooves that allowed an improved grip on the rod. The second was to wrap the handle area with rattan. I think rattan grips are really classy looking, but they are not ideal grips because they are hard and because a wet hand is just too slippery on the varnished surface. Still, up until about 1890 most fly rods had rattan wrapped grips. The shape of these earliest rattan grips was a small taper at the *cane end* and a gradual widening to the *reel seat end*. Simply put, a cigar shape. That would make the cigar shape the earliest and most logically named of all the grips.

As cork, in either sheet or rings, began to replace wooden and rattan grips as the norm, it was logical to continue the traditional shape. In 1885 Henry P. Wells book *Fly Rods and Fly Tackle* was published. In this book Wells strongly recommended a grip that was flared at both ends to *prevent slippage of the hand.*

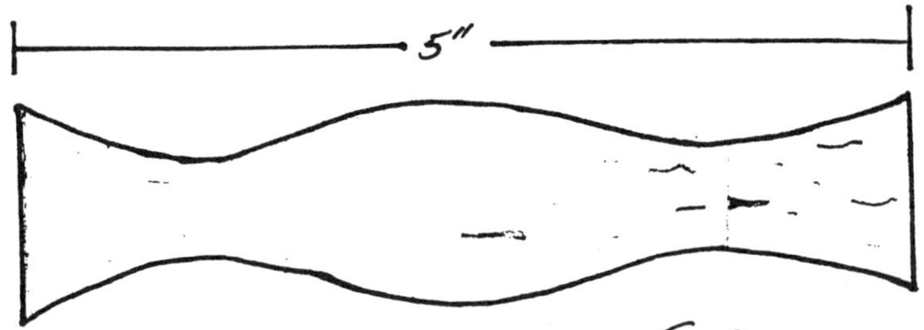

ORIGINAL WELLS GRIP

HALF-WELLS

cigar- straight taper

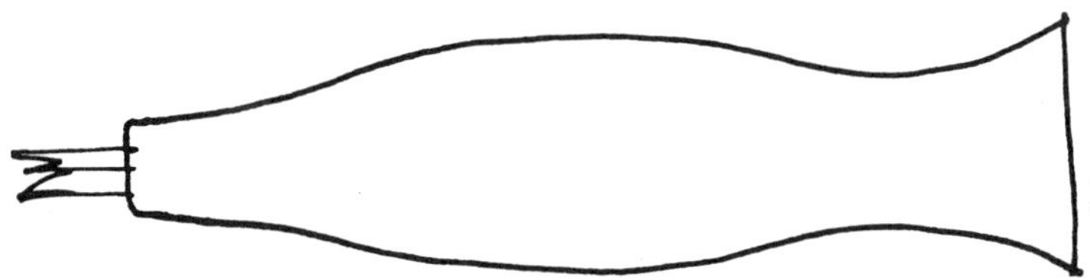

"COKE BOTTLE"

STRAIGHT (GORDON, EUROPEAN)

REVERSED HALF-WELLS

Ritz

cigar full taper

This is the famous grip that even today still bears his name.

In its original form, the grip's measurements were five inches long with a maximum diameter of one inch. While this seems to be too short for most of us, don't forget that it is an anthropological fact that folks were much smaller on the average in the 19th century than today. If you have trouble believing this just take a look at the uniforms worn by soldiers in the War for Southern Independence (that's the *Civil War* for you Yankees). Those uniforms look as if they were sized to be worn by children! The diminutive size of this grip is consistent with others of the period. Remember, earlier I mentioned that the shorter the grip, the older the rod.

Henry Wells would be proud to know that his grip has evolved into the shape most fishermen prefer today. In fact, other than the cigar and Ritz shapes (both basically straight), every other shape is some variation of the original Wells grip. Some of the descendants are the Half Wells, the Reversed Half Wells (usually called a Western), and Full Wells. Remember: Flared at both ends = Full Wells; flared at only one end = Half Wells. The standard Half Wells has the flared end nearest the cane while the reversed Half Wells has the flared end nearest the reel seat.

Sometimes, a grip seems to be a combination of two shapes. A good example of this is the current Orvis Superfine grip. It is a severely tapered cigar on the forward end and a half Wells on the aft end.

Other variations are simpler, just a slimming or fattening of the original Wells, or a slight relocation of the fatter mid portion of the Wells. For example, a typical Granger grip used with the patented uplocking reel seat is called a *mild reversed Half Wells*. The typical Heddon grip is generally referred to as a *long, very mild Half Wells*.

Even one of the truly original ideas for grips, the Phillipson *Hammer Handle*, is a form of the Wells grip. In fact the true name of the Hammer Handle was the *Elliptical Full Wells*. This grip actually has the shape of the handle on a hammer, and is quite comfortable to cast. Some of these fancy names are a bit pretentious

In many cases it would be more accurate to describe them in this manner: *I started with a Wells grip, then made it longer and slimmed it down some.* But somehow, that's not classy enough for catalog copywriters. They seemed determined to name every variation in shape that resulted from removing or adding a few millimeters to the diameter or length of a simple grip. That tradition is still alive and well today. Read the description, then look at the grip. It is bound to be straight, cigar or some form of Wells. Wanna bet?

In summary, the best way to identify an unmarked rod is to compare the features of the rod to known manufacturers' styles. Just as the police compare minute characteristics of fingerprints you can find the clues to the identity of the rodmaker from the components. I don't know precisely how many features must match on a fingerprint to make an arrest; but for matching a rod with its maker, I'd say three major points that match one maker should provide a pretty solid identification. For example, if the grip, reelseat and cane style (quality, color, nodes) all point to a specific maker, you've got a good idea who made the rod. You know the old saying, *if it walks like a duck, swims like a duck, and quacks like a duck, it's probably a duck.*

First Things First

Restoring a rod is twice the work of finishing a rod because you must *un-build* the rod before you can *re-build* the rod. Each step along the way to stripping a rod has potential hazards, and precautions should be taken not to make the condition of the rod any worse than it already is.

Before you do anything at all, read through all of the following chapters regarding un-building a rod and preparing it for restoration. By doing this, you will find the answers to many questions that will arise as you read. The text is laid out in the order in which the work should be performed in a step-by-step method. When there are choices of methods to be used, each possibility is discussed in turn and its good and bad points will be discussed to aid you in determining which method you want to try. None of these methods is foolproof, nor will all of them work every time with equal success. That's why options are discussed.

You may have varying degrees of success; experience and practice will help you learn or perfect the techniques. Speaking of practice – it is the best way to learn. Get some inexpensive *project rods* from dealers' lists or flea markets for your first attempts. It doesn't matter if they have one tip or two – or even if the parts are of equal length – you just need some cane rods to practice on that won't cause you to contemplate suicide if you mess them up. You can always refinish a rod then strip it and refinish it again. Try out the different methods – and some of your own as well. No one has all of the answers.

Bamboo Rod Nomenclature

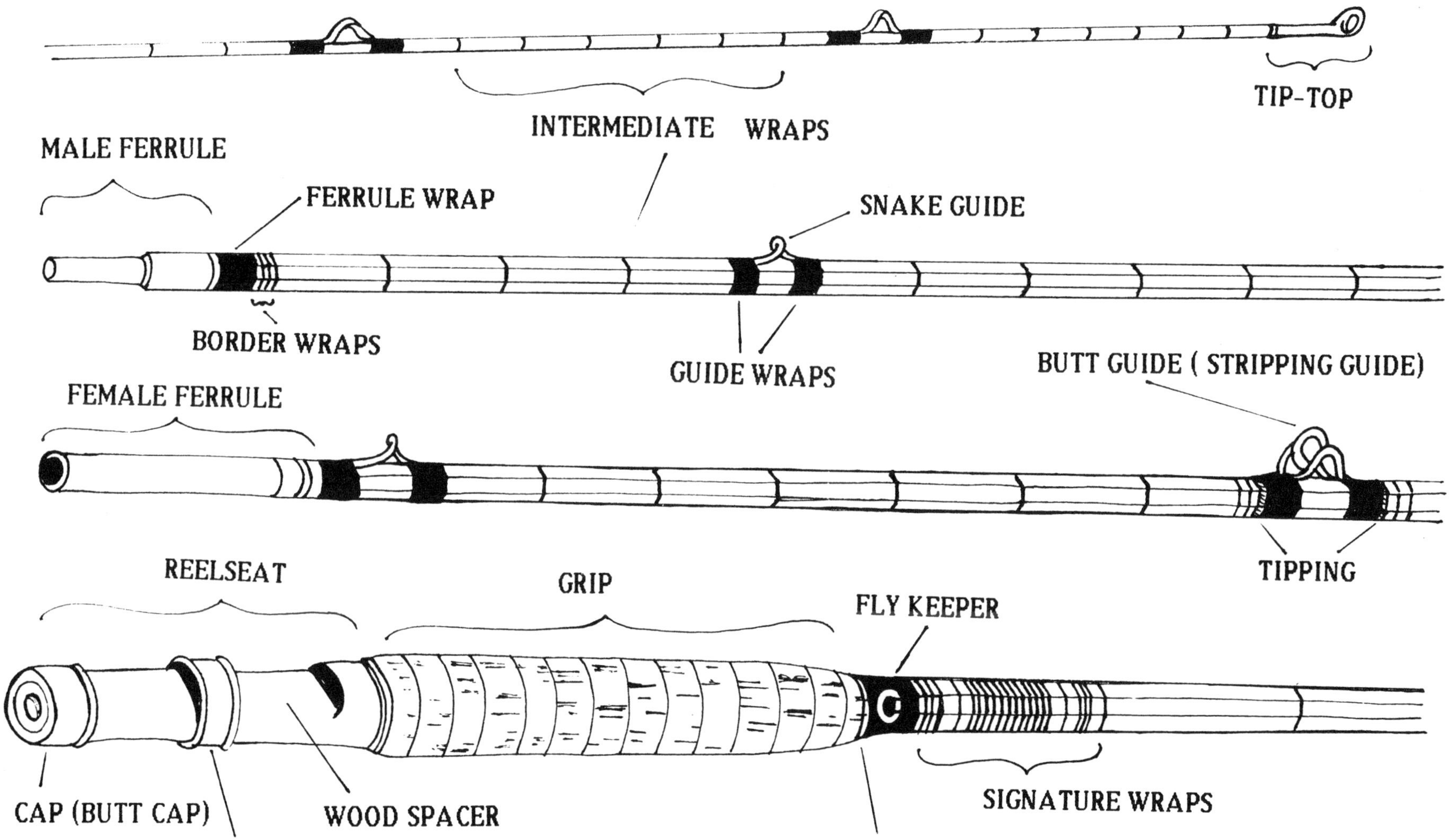

Logging In The Rod

Logging In the rod involves a complete physical description of the rod and its current condition. Your memory can play tricks on you, don't trust it.

Begin by completing a Rod Evaluation Sheet (See sample, page 29). This form can save time, and will also serve to remind you of the items you need to note on each rod. Make copies of the blank form on page 58, and keep a supply on hand.

Use the completed rod evaluation sheet sample as a guide for how to use the form. You'll need your yardstick and a set of calipers, a micrometer or a drill bit gauge for measuring guide spacings and ferrule sizes. Be sure to record all of the details as shown on the sheet, they are each there for a reason.

For the grip drawing, just lay the rod on the sheet and trace the grip. Mark the rings as shown for diameters at the center of each ring so that you may duplicate the grip if necessary. If you wish, you can draw the reelseat, ferrules and hardware on the reverse of the sheet so you'll have it as a reference. If you are a photography buff, use a macro lens or a telephoto lens to photograph details and an overall shot of the rod for later reference. It is interesting to have before & after shots of rods for your files. Even a Polaroid shot of the rod is better than nothing. Maintaining such a file of completed rod sheets and photos will help you when you come across a rod that is incomplete.

For accurate restorations it is also useful to *map* the rod. To accomplish this you will make a rough, but accurate drawing of the entire rod. This will be a full-size sketch to help you replace the wrappings in the exact locations where they were originally placed. This is especially important if the rod has intermediate or decorative wrap patterns between the guides.

Trace the rod onto either a piece of posterboard or several connected sheets of computer paper, usually 3 or 4 sheets is plenty. This is not an artistic effort; you simply make a vertical line as your starting point, and lay each section at the line. Draw a straight line the precise length of the rod section, then make a *hash-mark* at each wrap.

For guide wraps, put an X at the location of the guide itself, and note the size of the guide on your map. Make a double mark at the lower end of the guide foot so you will know the exact location of the guide. Guide wraps on most classic era rods were about 3/8" long, but some will be longer and others will be shorter. To preserve originality, mark the wraps carefully on your map so you can duplicate the original wrap length.

Save all of this information in a file folder. Before long, you'll have a good start on your own file of original specifications. For future reference, using the rod evaluation sheets, rod maps and photos, you'll know just how to restore that model.

MAKER: **MODEL:** **LENGTH:**

TIP FERRULE: / 64THS NUMBER OF GUIDES: ____

GUIDE SPACING MEASUREMENTS ARE FROM MALE FERRULE END

MID FERRULE: /64THS NUMBER OF GUIDES: ____

WRAP COLORS	GUIDES	TIP-TOPS	STRIPPER
MAJOR COLOR: ____	STYLE: ____	STYLE: ____	STYLE: ____
TIPPING COLOR: ____	COLOR: ____	COLOR: ____	COLOR: ____
INTERMEDIATE COLOR: ____	SIZE: ____	SIZE: ____	SIZE: ____
COLOR PRESERVED Y / N			MATERIAL: ____

WINDING CHECK: Y / N

HOOKKEEPER: Y / N

HOOKKEEPER WRAP: OPEN / CLOSED

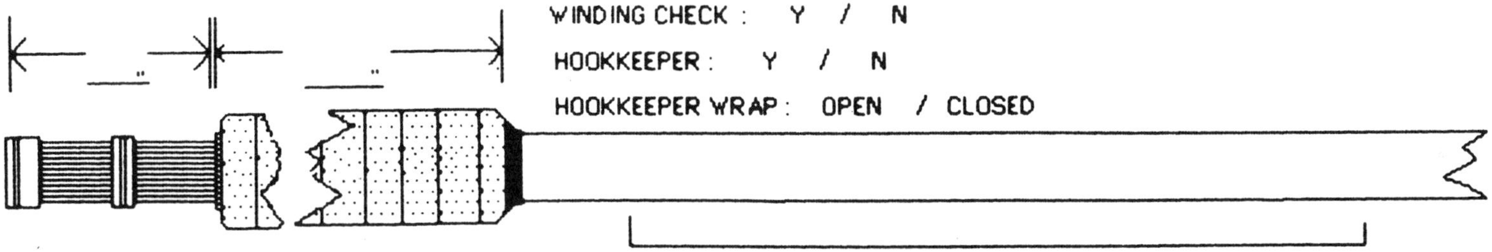

SIGNATURE OR BUTT WRAP PATTERN

REELSEAT	GRIP	CANE	FERRULES
STYLE: ____	SHAPE: ____	TYPE: ____	MATERIAL: ____
HARDWARE: ____	MATERIAL: ____	COLOR: ____	WELT STYLE: ____
SPACER: ____	WIDTH OF CORK RINGS: ____	NODE SPACING: ____	MACHINED: Y / N
COLOR: ____	# OF CORK RINGS: ____	VARNISHED / IMPREGNATED	SHOULDERED: Y / N
OXIDIZED: Y / N			OXIDIZED: Y / N

Tools and Supplies for Basic Rod Restoration

Tools:

Yardstick or tape measure
Pliers (with leather padded jaws)
4" Vise (with leather padded jaws)
Fly-tying scissors - Ice-tempered models stay sharp longer
Jeweler's or hobbyist's twist drill and bits
Jeweler's files (small needle files)
Gunsmith's punch with very fine points
Alcohol Lamp
Calipers or Micrometer
Good quality paint brushes – at least Gudebrod 1/4" quality, six
Better quality paint Brushes from art suppy house, 2 of each:
  Silver brand, size #4
  Sable Cat's Tongue, model #7110S
  Silver brand, size #7 Spotter, model #7517S
Cork gluing clamp
Thread tensioner

Supplies:

Pencil – Posterboard or computer paper
Scotch tape and masking tape (½" to ¾")
Razor Blades - single edge
NS wire – .040" and one size each way
Steelwool – 000 and 0000
Sandpaper- course through very fine, each grade
Wet/dry paper (400 and 600 grits)
3-M "Green Pads" for stripping varnish
Q-tips & cotton balls (synthetic is okay}
Ferrule cement (Hot glue)
Super Glue
White Glue – Elmer's or Sobo
Cork Bond glue by U40
Varnish remover – stripper
Denatured Alcohol
Acetone
Lacquer and Shellac
Lacquer Thinner
Mineral Spirits
Spar varnish
Chair-Loc
Facial Tissues - no scent or lotions! Cheap ones seem best.
Paper Towels
Brownell's polishes, 3F (Brownish Red) and 5F (Green)
Silk Thread
Simichrome Polish
Magic Wadding Polish
Mother's Chrome, Magnesium and Aluminum Polish
Water-based colored markers in various shades of yellow, brown, gray

Polishing Rods

". . . Varnish is a little rough, but will polish to excellent condition." How many times have you seen this statement in a seller's description? Polishing rods seems to create many questions. It is messy and time-consuming, but it is not at all difficult. Done properly, polishing can save 99.9% of the original condition of the rod and make it look close to new.

It is amazing how many rods that could have been polished end up being refinished unnecessarily. In some cases, rods with checked, alligatored, softened and sagging varnish can be polished and given a light overcoat of varnish to return them to a great looking finish without sacrificing originality.

Each polish job is an individual; what works once may or may not work on the next project, but don't be afraid to try polishing before refinishing. Try to polish a rod to save its finish if it does not need any other work such as guide replacement. If only the guide wraps are to be replaced, it is still possible to save the original varnish and ornamental wraps. The challenge is to match the original color of the wraps that will remain. It can be done. *See the chapter on wraps for details.*

One tip for now; if you will try to match the color of existing wraps, be sure they have been polished out before you try to match them. The color will be brighter after polishing, and that's the color you'll want to match.

The best polishing medium is the gunstock polishes from Brownells. Along with the polish, you'll need some 0000 steel wool, a generous supply of paper towels, some soft cloths and a worn out pair of pantyhose. You figure out how to get them. Brownells polish comes in two grades, 3F and 5F. The 3F polish is a kerosene based fine grit polish that is reddish-brown in color. This is the first polish you will use to remove the roughest part of the old varnish and help smooth out any nicks and scratches. Shake the polish before opening. It would be wise to spread some newspapers over the work area. Have your paper towels and steel wool close at hand.

Use a small piece of steel wool. Saturate the steel wool with 3F polish and begin to rub down the finish on the butt section, starting at the grip end. ***As long as you keep the steel wool saturated with polish and work gently, you will not damage the thread wraps. Use extra care on the corners; the thread can be damaged easily.***

Don't go at it as if you were trying to remove the varnish. Your goal is to have the kerosene in the polish soften the varnish so that you can redistribute it over the rod. Old varnish will soften easily if you take your time. It is not necessary to put much pressure on the rod as you rub it out. Work carefully around wraps and guides, and just let the polish and steel wool do the work. Don't be afraid to use too much polish. It is better to have too much than to let the steel wool get too dry.

Check the progress of your polish job by wiping off the excess with a paper towel. Change towels often, and when you've taken off most of the polish, let it sit for a few minutes to *dry* a bit. Then polish the varnish with a soft cloth, and you'll see what you have accomplished. If it needs more work, repeat the treatment. Probably, it will just need some touch-up in a few spots. Strips of material cut from the old pantyhose are great for removing polish that is stuck under snake guides. Slip one end through the guide, and clean it out with a *shoe-shine* motion.

When the rod has been polished thoroughly with the 3F polish, try using the 5F. This has a mineral spirits base. Apply this super fine polish to the finish with a soft cloth. Let it dry then buff it off with a clean, soft rag. The 5F will produce a very deep shine and will make the surface glass-like in appearance if all goes well.

In some cases, one polish seems to work better than the other due to the different bases. Experiment to see which works best on each rod. Rodmakers used different kinds of varnishes and finishes. You will find that it is virtually impossible to polish a polyurethane finish or one that was an oil finish such as Tung oil. Some of these can be saved with polish, but you will have varying rates of success.

Other polishing agents include pumice and rottenstone, which should be used with a light oil such as mineral oil and a felt pad. For a final, super fine polish, try powdered cornstarch and a soft cloth. You may discover that a combination of several different polishes will work best. As with anything connected with restoration work, practice helps more than anything else. Keep working at it and experimenting, and you can become very good at saving original finishes.

Saving Original Rodshaft Markings

A major area of concern in any restoration work is preserving the original markings on the rodshaft. These may be decals or hand-written script, which was usually done with carbon ink. When working around the area where these markings appear take your time and proceed with caution. It is most important to preserve all original decals or inked information. Some rods were actually autographed by the makers, and this is not only priceless, it is literally irreplaceable.

Determine the thickness and condition of the varnish covering the markings. If it appears to be thorough and in good condition, you may proceed to the next step. Most of the time, though, you'll need to do some work to preserve the markings. Even a partial decal is worth trying to save. Before going any farther, it is worth noting that sometimes, despite your most careful efforts, you just will not be able to save the decals and markings. The procedure outlined below has proven to be successful most of the time, and if followed accurately will give satisfactory results with regularity.

To save a decal that is dry, cracked and peeling and has little or no varnish covering it, use a fine-pointed brush to gently remove any dust or loose dirt on the decal. You must exercise caution or you may actually cause more damage. Don't make a bad situation worse. When clean, apply a tiny amount of fresh varnish. The best type of brush for accurately placing the varnish is found at an art supply house and is a #1 Spotter.

Your goal here is to get a tiny amount of varnish under the damaged decal to help secure it to the rodshaft. Treat each loose area this way. Next use a ¼" brush to place a thin coat of varnish over the decal. Let this varnish dry for at least a week. *Yes, a week.* After curing, the varnish will be hard enough to protect the decal when the rod is stripped.

To protect the markings or decals during the stripping procedure, cover them with a strip of paper and secure the paper to the rod shaft with masking tape. Be sure the tape is securely stuck to the cane to prevent stripper from seeping under it during the stripping procedure. This will leave a band of old varnish on the rod which can be dealt with at a later time.

Stripping the Rod

Once the rod's current condition has been accurately recorded, it is time to strip down the rod for restoration.

For the first part of this procedure you will need a container to hold the parts you remove from the rod. An empty 35mm film canister is a handy size for this job. An alternative for parts storage is a small zip-lock sandwich bag. A sandwich bag will provide room for ferrules and the reelseat if they are to be removed. Be sure to affix a label to the parts container that has the maker, length and model of the rod written on it to prevent mix-ups. Parts all look alike once they are off the rod.

Begin with the reelseat. If it needs work beyond polishing of the metal or is loose, remove the reelseat or the hardware. All-metal reelseats are easy to remove. Simply heat the reelseat over the alcohol lamp, grasp it with a hotpad, and pull it off. Just a note, do not *ever* use more heat than is absolutely necessary to accomplish your goal; all you want to do is soften the glue enough to release its hold. Be sure to rotate the work *above* the flame, not in the flame.

Take care that you do not burn or scorch the cork grip, or reelseat spacers of wood, cork or plastic composites. It should go without saying that you cannot heat wood, cork or plastic reelseat spacers with the alcohol lamp, but . . . *don't even think about it*. Some reelseats have a metal cap or barrel (spacer) that is pinned to the wood underneath. The pin is made of nickel silver wire and is usually about 3/8" long. To remove these pins, use a gunsmith's punch – about .039" is the most commonly needed size, though some rods will have pins of a larger diameter.

Remember to support the work firmly, and do not use too much force; the pin only needs to move about 1/64th of an inch to clear the thickness of the cap. Be sure not to mar or deface the hardware, be gentle. If you do this carefully, you will not enlarge the hole for the pin. This will simplify replacing the pin when you are done. After the hardware is removed, try to pull the pin from the wood with pliers and save it for reuse.

In the case of a pin that is damaged or cannot be removed from the wood, you can always make a replacement from nickel silver wire that is about .040" (18 ga.) in diameter. This is usually available from a jeweler's supply house. If the reelseat is of the type that has a wood spacer with metal threads at the grip end of the seat, you must leave the threaded part on the reelseat unless you plan to replace the wood spacer as well. It is nearly impossible to remove a wood reelseat spacer without causing it, or the rod, serious damage. If the threads are at the bottom of the reelseat, you may apply a little heat and try to remove the threaded portion of the barrel. Success rates vary.

When you have removed the reelseat, it is a good idea to weigh the rod on a postage scale for reference. The weight of the rod minus the reelseat was the weight used by most makers in their catalogs and advertising copy. Note this weight on your Rod Evaluation Sheet. Now it is time to begin removing the guides and other rodshaft hardware.

First, remove the guides. Begin by cutting the thread on top of the guide feet with a very sharp single-edge razor blade. This will help prevent marring the bamboo rodshaft. When the thread has been cut, you should be able to *scrape* the thread from the guide foot with your thumbnail. Usually the thread will come off in a rather large piece.

By examining the underside of this thread, the part that was next to the cane, you should be able to determine the original color of the thread. Often it is not even close to the color that showed on the rod. Varnish darkens and ambers with age, and this tends to mute the color of the thread. What appeared to be dull brown or dark gold may turn out to be a lively yellow or brilliant lime green. Save a piece of the guide wrap and tape it on the Rod Evaluation Sheet *bottom side out* for future reference. If there is more than one color be sure to save a sample of each. Even if you do not plan to restore *as original*, you'll know the correct colors if you ever need to restore a rod of that make and model.

As the guides are removed from the rodshaft, place them in the storage container. When you have removed all of the guides and their thread windings, remove the hookkeeper (if the rod was equipped with one) by cutting the wraps over the feet of the retaining strap. Don't forget to put both the strap and ring into the storage container. Remove the tip-tops by applying a little heat and pulling them off.

The last piece of hardware to remove is the winding check, or *taper*. This part is usually made of metal, normally nickel-silver or nickel-plated brass. Occasionally they were made of Bakelite or a plastic composite. The winding check was not glued to the rod, it was simply held in place by the silk winds. It is not necessary to use heat to remove these. Be sure that all of the varnish, lacquer and thread has been removed from the cane in front of the check. Slide a thumbnail under each side of the check on the grip side and *wiggle* it loose. Most will come off with a little encouragement, but some can be tough. Just stay with it, and be careful; these parts can be marred or bent out of shape very easily.

When the winding check begins to slide down the rod shaft take extreme care that it does not scrape off your preserved decals or rod shaft markings! Also move it carefully to prevent rounding of the corners (glue joints) on the cane. In rare circumstances, the check may fit so tightly that it cannot be removed without causing damage. If this seems to be the case, try to move it just enough to polish the metal without marring the cork grip. If you do remove the winding check, place it in the container with the guides and other hardware to be cleaned and polished.

The final step in stripping the rod is to remove the varnish. In most cases this is actually very easy. You have two goals in this procedure: One is to remove only the varnish, the other is ***not*** to remove any of the cane or mar the surface in any way. This includes protecting the sharp corners of the glue joints. *Rounding* of the hexagonal rod joint is a sure sign of an amateur refinish job.

There has been much discussion about the best procedure for removing varnish. Some feel that it is dangerous to use varnish stripping chemicals as they fear that it would harm the glue that binds the six strips of cane together. While that is a valid concern, if done properly and with the right stripper there is little chance of this occurring. The only alternative to using chemical methods is to scrape the varnish off of the rod.

This is potentially hazardous as it is very likely to cause some removal of the power fibers on the outside surface of the cane, and will in all probability cause damage to the corners of the hexagon. Some careless refinishers have been so thorough and diligent that they actually changed the taper of the rod by removing too much cane!

Here's the correct, safe procedure for using a chemical stripper to remove the varnish. Use 3M's *Safest Stripper*. This stripper has several qualities that make it ideal for rod restoration. First, it has no methylene chloride which is a dangerous and potentially hazardous substance that is common in most strippers. This product also has no foul odor, no harmful fumes and will not harm your skin. Additionally, it requires no neutralizing agent to stop its action and will clean up with soap and water. It is even safe if your cat or dog eats it, though that isn't recommended.

Even with all of these desirable traits, it works very well and relatively quickly. It will do a thorough job if given ample time to work. Some thin, older varnish finishes will be soft enough to remove in 45 minutes, but usually it is best to wait an hour or more. In extremely tough cases, the stripper can be left on overnight without drying out.

Along with this stripper, get some 3M *green pads*. These are like a synthetic steel wool that can be rinsed out and re-used many times. Steel wool should not be used with this stripper, as there is a chemical reaction that leaves a black deposit on the cane which is difficult to remove.

It is important to line up your work before you begin. You will need two pencils, some newspapers, a small dish or cup, and a wide paint brush (1½") to apply the stripper. It is also handy to have a sink nearby for wiping down the stripped rod and for cleanup when you are done. You will also need a generous supply of paper towels or rags you can throw away. When you've assembled your tools and spread out your newspapers, bring the rod to your work station. Space the pencils on the newspapers so that you can lay the ends of the rod joints on them and keep the joints elevated above the newspaper.

Pour some stripper from the jug into the cup for easier loading of the brush. Start with the butt section and apply a generous amount of stripper to all of the cane surfaces. It is better to have too much than not enough. When the butt is coated heavily, lay it across the pencils and repeat for the mid and tip sections.

While the stripper is working, you will have time to start soaking the guides and rod hardware for cleaning. To clean the old dried varnish from the guides and hardware, you'll need a small glass jar; a spice jar is just about right.

Place all of the guides, tip-tops and reelseat hardware in the jar and fill the jar with lacquer thinner. The hardware will need to soak for a few days to easily remove the varnish. It is a good idea to agitate the solution twice a day or so. Just swirl the solution around in the jar. Don't shake it. Shaking too vigorously could cause scratches and other damage to the hardware.

After the stripper has been on the rod for about an hour you can test to see if the varnish is softened enough to begin removal. To check, use the green pad to wipe away some of the stripper and, hopefully, the softened varnish. If the varnish comes away from the cane easily, it is time to finish stripping the rod. Use the green pad to wipe away the stripper and old varnish from the butt section. It should not require a great deal of pressure to remove the varnish, so don't put a lot of pressure on the pad. Even this pad could cause some damage to the corners of the cane if you get careless.

When working on tip sections, be very careful to support the section while wiping. It is best to lay the tip on the newspaper so that it is entirely supported and cannot be bent. You may notice that there will be a few small areas where the varnish seems unaffected by the stripper. Don't worry about these spots now; just remove what will come off easily. When the pad becomes heavily saturated with the stripper/varnish combination, rinse it in warm water, wring it out and continue to remove varnish in this manner until the section is clean. When all varnish and stripper is wiped from the rod, use a dry paper towel to wipe more of the residue from the cane. It may be necessary to do this a couple of times as the stripper and varnish solution is gelatinous and sticky.

Next, wipe down the section with a damp, *not wet*, paper towel and set it aside to dry thoroughly. While the cane is air drying, you can clean up your mess.

Once the cane is dry, it is time to touch-up those stubborn varnish spots and finish cleaning off the blank. For this part of the job you will need a can of acetone and a good supply of cotton balls. It is fine to use synthetic cotton balls or cotton pads. Soak a cotton ball with some acetone and wipe the blank. When you reach one of the varnish areas, soak it with acetone, and go on wiping the rest of the blank. The cotton balls will get dirty quickly, so you'll want to change them often. They work better when they are clean. When you have finished with each section, go back to the varnish spots and scrape the varnish with your thumbnail. It should flake off fairly easily, but if it does not come off cleanly, wipe it again with acetone and repeat scraping with your thumbnail until the cane is clean.

Now is the time to remove the masking tape that you placed on the rodshaft to protect the markings. Use the acetone and cotton balls to clean away as much of the old varnish as you can without getting any acetone on the marking areas. Acetone and varnish stripper will dissolve the decals and remove the inked markings from the rod, so be careful! Clean each section this way and take extra care and caution when working on the fine tip sections. If you've been thorough and followed the instructions above, you should now have a perfectly clean blank with sharp corners; just like the rod looked when it was in the factory.

Refurbishing Ferrules

Ferrules are critical components for any bamboo rod. They were either the rodmaker's pride, or the bane of his existence. A fine ferrule design and execution is the most important factor in determining the grade of a rod's quality. If not for poor quality ferrules, many rods from production companies would rival their higher-priced cousins on a functional basis. High quality ferrules not only help performance of the rod in casting, but can actually help the rod last longer.

Ferrules that are serrated help protect the cane at the critical stress area where the energy of casting is transferred from the metal ferrule to the bamboo. Basically, serrations are of two types, one is the *Split Style* serration as patented by H. L. Leonard, the other is the *Pie-Wedge Style* serration as patented by George Varney. Split-style serrations are actually sawn into the wrapping shoulder, where the Pie-Wedge style serrations are usually filed or punched.

Waterproof ferrules prevent moisture damage from entering the end of the rod. Ferrules that have a close tolerance friction fit provide an even, smooth transfer of energy from one rod section to another, and the tight fit prolongs the life of the ferrule itself. Only a few makers made very high quality ferrules. The material of choice for high quality ferrules (and for all metal parts of a rod) has always been nickel silver. Remember that nickel silver is not really silver of any kind; it is yellow brass (an alloy of copper and tin) with nickel added (18 parts per 100, or 18%). That means that the proper name for nickel silver is *white brass*. Another common name for this alloy is German silver, though it was also called Liberty Silver during the period of World War I.

Usually, the best ferrules are referred to as *machined*, though this is not entirely accurate. These ferrules were actually made of two different diameters of nickel silver tubing, and had a waterproof seal soldered on the end of the male and at the bamboo seating location inside the female. In removing such ferrules it is important not to use too much heat, as you can melt the solder that holds these discs and the welt in place and damage the integrity of the ferrule.

Many of these ferrules were also pinned in place. Exercise much care in removing the pins. You do not want to mar the metal work or deface the finish of the ferrule. Rods that used the highest quality ferrules were those such as Payne, Leonard, Hawes, Gillum and a few other fine craftsmen.

Cheaper ferrules were called *drawn* ferrules, as they were drawn from a sheet of nickel silver. These can have the same protective features such as waterproofing and serrations, but they do not have the same precise fit. Though entirely adequate for fishing purposes, they are simply not as good.

Lesser materials were used for ferrules as well. These were usually some type of plated yellow brass, most common being nickel plated brass. Some makers of production rods used chrome plated brass, and some used yellow brass that was oxidized. The ferrules used on many antique rods were made of nickel silver sheet metal that was rolled into a tube and soldered.

Better ones were then improved by the addition of a welt, a ring of metal, at the mouth of the ferrule to help it hold its round shape and to prevent damage. Because these were soldered, you must exercise caution in heating them for removal, and remember to check for pins. It may not be necessary to remove these ferrules at all. Antiques are not fishable by today's standards, though they can be fun. If the rod won't be fished, leave the ferrules alone; especially those of spiked or doweled design. You have heard of Pandora's Box, haven't you?

The shape and style of the ferrule and its method of manufacture are good indicators of both the maker and the grade of the rod upon which they were installed.

From an engineering aspect, some shapes are more efficient than others. The most efficient is the design originally called the "Super Z" ferrule, which was patented in 1952 by Louis Feierabend. Modern variations of these ferrules are called Swiss Style. Perhaps the most ardent supporter of Super Z ferrules was E. E. Garrison. Being an engineer, he immediately recognized the superiority of its design which allows the diameter of the cane to be the same on the two sections that will be joined.

The least efficient ferrules were those with a *dual diameter*, which gave them the look of a shouldered red wine bottle. The only good point about these ferrules was that they required very little removal of the outer fibers of the cane, but their weakness was that they were not serrated, nor were they wrapped over a winding shoulder to seal the joint and strengthen it. These ferrules were usually pinned. These were made of different materials, but not many were nickel silver. If you see these ferrules on a fly rod, it is probably a Montague Rod, and the ferrules are probably nickel or chrome plated brass. Several manufacturers used this bottle shaped ferrule on bait and casting rods.

Many ferrules fall between these quality levels. Good examples of a *functional* ferrule were those used on rods by Granger and Phillipson. Though these were made of *hard drawn* nickel silver, they were otherwise pretty average. They lasted better than many others, but the fit still was not as good as the *machined* variety. These ferrules are interesting because they did not have serrations, but they were feathered in thickness of the wrapping shoulder to accomplish the same purpose as serrations. This feathered edge was shaped to fit the cane after the ferrules were mounted to the rod. Phillipson and Granger ferrules were not pinned, but seldom work loose from the rod.

When restoring a rod, you must decide the purpose for which the rod is intended. If it is to be a part of a collection, leave the original ferrules for authenticity. If the rod is to be fished, and the ferrules are not up to the task, you may want to replace them with a modern alternative.

The finest ferrules available are from Bailey Woods' Classic Sporting Enterprises. They are called Super Swiss and Super Step-Down. The step-down ferrules are of particular interest to restorers as they are the style of the better ferrules used on most classic rods. The Dual-Diameter (Montague) style is also available for those rods that require them. Woods will copy classic ferrules for accurate restoration of fine rods, but it is best to write for information and to see what the cost will be. In most cases, only the very finest rods would justify such expense.

Of course, you can always check with other restorers; you may get lucky and be able to trade for the part you need, or purchase it outright. Keeping the original ferrules is the best policy if at all possible. If the fit is loose, which is detectable in the form of a *click* in the joints when casting, you must first determine which ferrule is loose. If it is a three-piece rod, assemble only the mid and butt sections, and flex the rod as in casting. Try to feel or hear the click. If it is present, you know you'll need to work on these ferrules. If you don't feel the click this way, assemble only the mid and tip and flex again. Continue with different combinations until you have determined which ferrules have a problem.

Just a word of advice – if the fit is loose between the metal and the cane on one section, it will probably be loose on all of the sections. That's a function of the age of the glue and the way the rod has been stored over the years, and should be addressed equally on each ferrule. The solution is easy. Remove the ferrule, clean the old glue from the cane and inside of the ferrule with acetone and a Q-tip, apply new glue and reinstall the ferrule. If the ferrule was originally pinned, you'll want to replace the pin as well.

If the problem is that the male ferrule is too loose in the female, you face quite a challenge in correcting the problem, but it can be done. For one method, you will need access to a lathe with a three-jaw chuck. This chuck will allow you to place equal pressure on the outside of the female ferrule to attempt to compress the opening and make the fit of the male more snug. It is best to insert the male ferrule into the female and be sure it is lined up as for regular fishing. You absolutely must exercise caution in this operation, or you will ruin the ferrules completely.

Use masking tape to cover the female ferrule and help protect its finish. Place the assembled ferrules in the three-jaw chuck and tighten the jaws very slightly to apply pressure. Rotate the female ferrule and repeat a few times, attempting to compress *all sides* equally. This operation does not require that you change the ferrule very much; take your time and frequently check the new fit. If you are working on the female that accepts the tip ferrules, check both tip ferrules for proper fit.

Another way to accomplish this tightening requires no machinery. You'll need two large pieces of very, very soft lead. Make an *anvil* of lead about ½" thick and four inches square. Use the second piece of lead to make a *club* in the shape of a baseball bat. Lay the assembled ferrule on the anvil, strike it with the club, rotate and repeat, checking the fit between strikes. If you are careful, you'll have very satisfactory results. The soft lead will not mar the surface of the female ferrule, and usually the ferrule will show no sign that it has been adjusted in this manner.

Re-seating ferrules is not at all difficult. When you removed the ferrule, you may have noticed a piece of string or thread folded over the end of the bamboo. This means that the cane was turned down too far at the factory, making the ferrule station too small. The thread was used to increase the diameter of the ferrule station in an attempt to make the ferrule fit tightly. If that is the case, you must replace the string when you re-glue the ferrule. It is always best to clean away as much of the old glue from the ferrule station and from the inside of the ferrule as you possibly can. It is not a bad idea to *rough up* the surfaces a little before applying the glue. When you reinstall the ferrules, pay attention to keeping the witness marks (if any) lined up with the spline of the rod.

Check the spline of the rod yourself; the factories often paid no attention at all to the spline and you can improve the performance of the rod if you take the time to spline it properly. Innovative rodmaker Letcher Lambuth called this *tuning the rod*. He tested many factory rods from low to very high quality and found that only a few were properly splined, or tuned. His tests indicated that an improvement of as much as 25% was possible.

Once you have determined the spline of the rod, be sure to place the witness marks on the proper flat. If there are no witness marks on the ferrule, mark the #1, or guide flat as determined by your splining, with a permanent marker by making a small black dot on that flat right at the edge of the ferrule where it will later be covered by a wrap.

Ferrules that were pinned will need to have a new hole drilled for re-pinning if you rotate the ferrule from its original position. This should be done cautiously, as the second hole in the bamboo may weaken the cane excessively. It is your decision to make. If the ferrules were oxidized or Japanned, now is the time to recreate the original finish. To *blue* or oxidize 18% nickel silver, the easiest method is to use Birchwood-Casey Brass Black. This product is available at most gunshops, and through Brownell's.

Since nickel silver is *white brass*, Brass Black works very well. Using a treatment that affects the nickel in the alloy can be frustrating. If the nickel silver is less than 18% nickel (usually 10% or 12%), you will have great difficulty in getting it to darken, and the results will probably be blotchy. Other popular gun bluing treatments may work, but some require heat. Experiment until you find which works best for you.

Kirkfield's book has a recipe for an authentic oxidizing agent. If you are a chemist, or know one who will help make up the potion, it is extremely effective and produces a deep blue-black oxidation. However, this is a potent chemical mixture and its use presents many potential hazards. Use it with caution.

Whatever method of oxidizing you choose, it will be essential to clean the metal thoroughly. Use 0000 steel wool for most of the cleaning. For stubborn tarnish or on pitted areas, you may want to use 400 grit wet or dry sandpaper followed by the steel wool. After this cleaning, use denatured alcohol or acetone to wipe the metal clean and remove any traces of body oils from handling. Once clean, don't touch the metal with your hands. Then follow the directions for the oxidizer you choose.

The reaction sometimes takes five minutes or more; don't get discouraged. If you have cleaned the metal parts thoroughly, you should end up with a good, even blue-black color. After the oxidizer has been removed, buff the darkened ferrules with a tissue. A coat of spray lacquer will help prolong the life of the oxidized surface. Keep the lacquer off of the cane! If you don't seal it, the blue-black may fade to a dark brown; not unattractive, but not as classy as the deep black finish.

An alternative to oxidizing is Japanning. This is a process that is essentially painting of the metal with color-tinted lacquer and then sealing it with a coat of clear lacquer. Heddon ferrules were usually Japanned, though the ones they called Gunmetal Finish were actually oxidized. This method is best for metals other than nickel silver or brass, such as aluminum, which was widely used for screwlocking nuts on reelseats. You can buy Japan colors to customize rods in some interesting ways, including giving a gold color to the ferrules as on the Heddon #125 Expert model.

For black ferrules, the easiest method is to paint the ferrule with India ink on a Q-tip, and when dry, seal it with a thin coat of polyurethane. The polyurethane will dry quickly to a very hard protective coat. The same methods may be used to darken reelseat hardware.

Cleaning, Repairing or Replacing Cork Grips

A good place to begin is to determine whether you want to do anything at all to the grip. There really is no harm in leaving the grip in its *dirty* condition; it shows the rod to be a veteran of the stream. The old brown coating is formed with use over the years and is composed of sweat, dirt, line dressings, insect repellents and fish slime. Some people prefer to leave them in their current condition.

If you are performing a restoration for someone, it is wise to ask their preference. There is a good reason for this; if the grip is cleaned you run the risk of having the old cork deteriorate. This is usually noted in the formation of ridges caused by sanding or otherwise cleaning the grip.

These ridges are formed in two ways. First is because the cork is softer between the ridges and some of the softer areas may be removed simply by removing the brown coating that accumulates on grips. Another cause of ridging is that all cleaning methods remove some of the natural moisture from the cork and make it a bit brittle. This is not to say that there is always damage done to a grip by cleaning, but you need to be aware of the potential risk. The age of the cork is an important factor to consider. Remember that even the *later* bamboo rods are over fifty years old today.

If you determine that the grip must be cleaned, use a mild detergent such as Amway's Liquid Organic Cleaner (L.O.C.). This detergent seems to do the best job of removing the brown coating without causing unnecessary desiccation of the cork. Use it full strength, and rub it into the grip with your fingers; don't use anything abrasive. If your fingers aren't rough enough to do the job, go to a gourmet shop and buy a *mushroom brush*. This brush has soft bristles that will not damage the grip as it is designed to clean fragile mushrooms without damaging the skin. After brushing or rubbing the grip, rinse it under warm water, taking care not to wet the cane excessively.

Moisture is one of a bamboo rod's worst enemies, so you don't want to overdo it. Blot the grip dry with a paper towel; don't rub it. Let it dry for an hour or so, and evaluate whether it is now clean enough. If this system does not seem to do the job for you, wipe the grip with cotton balls and acetone. This will literally melt away the old brown coating, but it does tend to dry out the cork a bit. Use it as a last alternative for cleaning cork.

Cork grips may be sanded to remove the old coating or to remove ridges, but caution should be exercised not to remove so much cork that the grip shape is changed. Use the finest sandpaper that will do the job. If the sanded cork seems to want to form ridges, you may want to replace some of the moisture by applying some baby oil. To use the baby oil, apply a small amount of the oil to your hands, rub it in, and then rub your hands over the grip. If you don't go overboard with this treatment, it will not make the grip slippery. It will, however help prevent formation of ridges.

Minor dents and gouges in cork grips can be fixed by holding the grip over steam from a teakettle. The steam adds moisture and heat which helps the dents pop back out. It won't work every time, but it is worth a try.

Repairing hook damage and other cuts, holes or gouges in the grip is easy. There are two fillers that are commonly used. The most effective and easy to obtain is a mixture of cork dust and lacquer (or cork dust and white glue such as Elmer's or SOBO). This is the best filler, as the cork dust will blend better with the color of the grip. If the color is not just right, you can darken the filler by staining the cork dust with left over coffee or tea.

If you are working on an uncleaned grip, shoe polish or Old English furniture polish may be used to darken the filler. Always overfill the hole just a little, but don't leave yourself a giant sanding chore. An alternate filler is Plastic Wood, or the equivalent. This filler is available in many colors, so you can match most grip or cork colors right out of the can. Plastic wood dries very hard, and can be sanded. Don't use this filler for large areas, as it is brittle and can make a grip uncomfortable for fishing as it is not flexible at all.

A grip that is too damaged for repair may simply be replaced. This is not really a major project if the rod must be stripped for restoration. Before you remove the old grip, trace its outline on a piece of paper. Note the location of the joints between cork rings, and the width of the cork rings. Be sure to measure the diameter of each ring at its center using your calipers, and record this on the sketch. Now you have an accurate blueprint for reproducing the grip as original. If the cork rings are any width other than ½" you will have some extra chores to do before you can recreate the original grip. Rings of over ½" on a rod indicate that the rod was probably very inexpensive, and has little collectible value. You are actually improving the rod as a fishing tool by using the higher grade ½" cork rings.

Just remember that you will need enough rings to match the original length of the grip. If the corks have a width of less than ½", they will be either 3/8", 1/4", or

1/8". In most cases, rods with *skinny corks* are higher grade rods. Before you take on the task of replacing such a grip, you really should consider what the replacement will do to the value of the rod, both historically and collectibly. Think long and hard before determining that replacement of this grip is absolutely necessary. If you must replace the grip, you should cut down ½" cork rings to the proper width and reconstruct the grip as original. This can be costly.

A six-inch grip of ½" corks will require twelve new rings, whereas a six-inch grip of 3/8" corks will need about sixteen rings. If you get down to the 1/8" corks, you will need about 50 of these for your six-inch grip! That's a lot of work, true, but a rod that had these very narrow corks is worth the time and expense to restore properly.

Cork rings of 3/8" can be made from ½" corks simply by sanding on a belt sander to reduce the width. That's a bit wasteful, but is unavoidable. If you are very careful, you can use a bandsaw to cut each ½" wide cork ring in half. Except for one half the width of the sawblade, you'll have two ¼" corks. That should be close enough. Unfortunately, to create 1/8" rings, you can only get three from each ½" cork, and you must then lightly sand them to the proper thickness.

A few words about the quality of the new cork rings is in order. The highest quality rings are sold as *flor quality*, sometimes called *flower*. Other terms used by cork sellers may be AAAA, or Superfine. These corks are usually of a higher quality than was originally used by rodmakers. Unless you are restoring an extremely high quality rod, it simply is not necessary to use these. Grade AAA is just fine for most restorations, and even a lower grade is suitable for most production rods, especially the lower grades. The less expensive corks will do just fine for making rings of 1/4" or 1/8".

Rebuilding a cork grip is not difficult, but it does present a few problems. First, you will need access to a lathe for turning the shape of the grip and keeping it *true*. The lathe should have a hole through the spindle large enough to admit the butt section of the rod for best results. If it does, you simply glue the cork rings directly to the rodshaft in the proper location and when dry, turn the grip to the dimensions you recorded before you removed the original grip. When you glue the corks, whether on the rod butt or as an independent unit, use a glue such as Cork Bond from U-40, or Tite-Bond woodworkers glue. The same white glue used for making cork filler will work as well.

Get a cork gluing clamp from a rod supply house, and clamp the freshly glued grip section for at least an hour, but it is not necessary to overstress and compress the corks; just tighten the clamp until the corks are snugly joined and the excess glue is forced out of the joints. Don't worry about this glue, it will sand off easily when dry. If you are using a wood working lathe, you will need to chuck the reelseat end of the rod in the 3-jaw chuck, support the other end in a roller bearing device, and then you may turn the grip. A very well made and inexpensive roller bearing rod support is available through Dale Clemens' rod making supply house.

Some rodmakers prefer to build the grip on a mandrel and glue the completed grip to the rod. This is a good idea for a couple of reasons. The best of these is that turning the grip on the rod butt is only effective if the butt is perfectly straight. Another reason is that turning the rod butt in the lathe is potentially harmful. The only drawback to this method is that you must bore the proper size holes in the cork rings before assembling the grip for gluing, and if the holes are smaller at the front of the grip, be certain that you keep the rings in the proper order when gluing up the grip.

Whatever method you use, take your time in turning down the grip and measure at each cork joint often to be sure that you end up with a grip that matches the appearance of the original. When turning, go through the grades of sandpaper (course to fine) to achieve a very smooth, velvety feeling grip. Don't go beyond about 150 grit, it won't look original. Be sure to use a tack cloth to remove all of the cork dust. This dust will make your life miserable when time for wrapping the rod with silk thread and when varnishing.

Restoring a grip that is *cane wound*, or *rattan wrapped* is very much like working on the cane itself. As with any grip, you may just want to leave it alone, or perhaps give the varnish coating a polish. If the grip needs to be refinished, you may remove the varnish from the grip in the same manner as used for the stripping of the rod, but you will need a disposable, stiff toothbrush for removing the varnish between the winds of cane. There is probably a winding of thread between the ridges of the rattan. Usually this thread was red, though occasionally you'll see dark brown or black. If the thread needs to be replaced, it is easy. Just wrap it from bottom of the grip to the top and the end should be covered by the winding check. If there is no check, this thread usually formed a decorative winding at the front end of the grip.

Rattan wrapped grips may be completely replaced, if you like a little adventure. The cane is called *round reed*, and is available in several sizes at large hobby stores that sell basket making supplies and at stores that cater to chair caners, or you can order this reed from Constantine's in New York. The size you'll want is 2.25 mm. The reed comes in a coil of several strands of varying lengths. Use the longest ones for your grips. A six-inch grip may require as much as nine feet of reed. If you have only shorter pieces, you may have to splice two sections together to make the grip. Do this before you begin. Once you've started you'll have your hands too busy to worry about it. It isn't all that hard to do, but it is a little tricky.

The reed must be soaked in warm or hot water for about half an hour before use. It will remain pliable for an hour or so. While the reed is soaking, remove the reelseat from the rod; it is probably all metal and easy to heat and remove. Note the location where the grip begins under the reelseat. There should be a place where the reed is inserted into a hole drilled into the shaft of the rod. This was the beginning point, so it is where you'll need to begin as well.

Remove the old reed by whatever method works best for you. It is probably held on mostly by the old varnish and crud that built up over the years. If there is a winding check, you'll want to slide it forward out of the way or remove it completely. When the old grip is removed, be sure to clear out the hole in the shaft so you can insert the new reed into it. After inserting the soaked reed into the hole, begin wrapping the reed around the grip area. Take care that each wrap lays tight to the previous wrap so you'll end up with a professional job. When the wrapping is complete, use some Tite-Bond glue and clamp it by wrapping with a large, very wide rubber band until dry. Again, this is tricky, but it can be done.

Set the rod aside to dry for a day or two. The last step is to stain the reed to a honey color to resemble the aged varnish. For this, use Minwax Golden Oak stain or Honey Pine. This stain will add a rich, aged color and help seal the grip. Usually it takes about three coats with a day between coats to dry. After the staining is complete, wrap the thread between the ridges and give the grip two coats of varnish. You may need to taper down the area that is covered by the winding check and the open end of the reelseat. A file and sandpaper are handy for this chore.

Sheet cork grips are difficult to replace only because the sheet cork is very difficult to find. If you must replace one use rattan wrapping or cork rings. Either would actually be an improvement, one for fishing the other for looks.

Reconditioning Reelseats

The easiest reelseats to work with are those made entirely of metal. The metal is likely to be nickel silver, nickel-plated brass or chrome plated brass. All of these can be easily polished, but those made of nickel-plated brass (NPB) require a little caution in polishing. Often the plating was very thin, and may even be partly worn away showing the yellow color of brass underneath. These reelseats are best cleaned with soap and water. After drying, they can be polished with *Magic Wadding* metal polish (from hardware stores) or Mother's Chrome, Mag and Aluminum polish (from auto parts stores and truckstops). Both of these are non-abrasive polishes that will clean the nickel without removing it. Chrome plated brass (CPB) seats can be cleaned in the same way using the same polishes with very good results.

Nickel silver reelseats are the easiest to clean and polish since they are made of a solid alloy rather than being plated. No matter how bad a nickel silver reelseat looks from corrosion and minor scratches, it can be made to look as good or better than new. For N/S reelseats with corrosion or heavy tarnish, begin by cleaning it as if you were polishing a rod; use the Brownell's 3F polish and a bit of 0000 steel wool and gently wipe down the entire reelseat. You will note that the polish helps break down the corrosion and darkened areas will begin to look clean. When most of the heavy deposits have been removed, wipe the reelseat clean with a tissue or paper towel. Repeat if necessary. This will usually be all of the cleaning that is needed, and a little buffing with a tissue will make the reelseat shine beautifully. If there are nicks and scratches, try rubbing them out with 0000 steel wool, then polish again. If you wish a very shiny finish on the nickel silver, use a MotoTool with a buffing pad and some jeweler's polish called ZAM. This will make the nickel silver look like highly polished Sterling. Some people do not like this extra polishing (remember to ask for the preferences of the rod's owner). The same cleaning methods listed above may be used for ferrules of various metals.

Reelseats that have wood spacers present a whole different set of problems. If the reelseat is a simple cap & ring style, it is usually simple enough to remove the butt cap with a little heat. Watch for it to be pinned in place; most of these were pinned. Remove the pin with a gunsmith's punch and try to save it for reuse. Some reelseats have a very narrow band of metal between the wood and the cork grip. This is called a reelseat check or a grip check. Most of these have a pin in them as well, and can usually be removed when the pin is taken out. When all of the hardware is removed, it is a simple matter to strip the varnish from the reelseat spacer and apply a new thin coat of varnish.

Don't use much varnish. The classic rodmakers did not finish reelseat spacers with a high-gloss, high-build finish. They usually just put a little varnish on the wood and rubbed it in thoroughly. Use your finger for this, it will do just fine. Cover the wood with the varnish, but very, very lightly. If the wood spacer is damaged by gouges, or has cracks from stress, you may want to consider replacing it. If it is a simple design (i.e.: not mortised) you may want to replace it with a modern wood spacer.

If you do, remember to try to match the original wood and finish. A good selection of spacers is available through rodmakers supply houses and through REC. If you have access to a wood working lathe, consider turning your own spacers. It isn't hard, and makes the final result more satisfying. A milling attachment for the lathe or a router is needed to add the mortised groove to a reelseat such as seen on Divine Rods. This is accomplished by using a 1/2" Fingernail Bit, available through Cascade Tools.

Celluloid was a popular reelseat spacer material in the era from 1890 to the

1920's when it was replaced by hard rubber and later by composite plastics. It is now illegal to make celluloid due to some of the hazardous materials it requires. If you have a reelseat spacer of celluloid, it cannot be replaced with the same material. A rod that has a celluloid spacer may be too collectible to change its specs.

A good example is the ultra-classic Kosmic Rods. For one period of their production, Kosmic Rods had a white celluloid spacer called *Ivoroid*, which was a substitute for ivory. It is difficult to discern the difference between black celluloid and hard rubber just by looking. Here's a way to tell. Polish a small part of the material with 0000 steel wool and smell the material. Yes, I said smell it! If you detect an odor of camphor, it is celluloid. If it smells like burned rubber . . . well, there you are. If you wish to save a celluloid or hard rubber reelseat, it must be polished while installed on the rod. You may be able to remove a butt cap or other hardware, but usually not.

Reelseats made of Bakelite, Pyralin, and other plastic composites are the toughest to refurbish. If these are cracked, or excessively scratched or beat up, they are not worth saving. In just about every case, any rod that has this type of seat is common enough that a collector can afford to wait for a better example for his collection. No one else would care if these were original. Though some people think of plastic reelseat spacers as cheap and tawdry, try to remember that when they were introduced they were the new technology, and actually were an improvement. They did not require varnish finishing, did not swell when wet, and did not crack as easily. They are, however, a real pain to work with if they must be removed.

There is only one way that may work to heat them for removal and it only works about 50% of the time. The reelseat (still mounted on the rod) must be placed in a small pot of boiling water. Hopefully, this will heat the glue enough to allow the reelseat to be pulled off the rod. Sometimes this boiling will help remove minor dents from the plastic, but this depends upon the type of plastic that was used.

The problem is that with many materials, the heat will cause a lightening of the material. This is especially true with black Pyralin. The lightening may be corrected by polishing the reelseat barrel with Simichrome Polish and then buffing by hand. This polish is also effective on most plastic reelseats that have lightened up due to drying out and sunlight fading. It does require a lot of elbow grease, but the results can be very good.

If you have decided that a reelseat spacer of wood or plastic must be removed and the glue just won't let go, the only option left is to use a chisel to remove it. Be very careful not to cut the bamboo where it passes through the reelseat! When replacing a reelseat with a new one, be sure to order a reelseat spacer that is bored to the correct inside diameter to receive the insert. Inserts can be turned or sanded to a smaller dimension if necessary, but if working with the butt end of the cane itself, don't remove too much or you may weaken the cane and cause the reelseat to break off. If you must increase the diameter of the insert by just a little, wrap it with thread to make it slightly larger. In some cases, you'll need to enlarge it quite a bit. Use reject quality cork rings for this and turn them to the necessary diameter.

Good quality reelseats of modern manufacture may be found through all of the rodmakers' supply houses. The better reelseats are made by Struble and by Bellinger. The highest quality seats are available through REC, which offers some nicely knurled classic styled N/S hardware and mortised spacers in a very wide variety of fine woods. A very interesting reelseat is available through Belvoirdale. The hardware is made of Titanium or nickel silver, and they offer a unique *upsliding* cap and ring seat. The choices are yours to make; but you will find that the one thing that has not changed in components for over a hundred years is the wide variety of prices for different qualities of reelseats.

Salvaging Guides and Tip-tops

Many of the guides and tip-tops used by classic rodmakers have become minor collectibles. They are, of course, no longer manufactured, and the style of the current guides doesn't usually match the originals. If a rod is destined to be part of a collection, it may be best to re-use the original guides when the rod is restored. If the old guides are not to be reused, save them anyway. One day you'll be glad you have them, even if they are not in perfect condition.

It always seems that the rods that you want to leave all original are missing a guide or two, or have one that is broken. Saving the guides was briefly discussed in the section on stripping the rod. If you followed those instructions, the guides and tip-tops are now in a glass jar soaking in lacquer thinner.

We'll begin again at that point. After a day or so in the lacquer thinner, most of the old cruddy varnish will have softened, and much of it has gone into solution. The remaining varnish is usually very easy to clean away. You will need a measuring cup with a pouring lip, a small (2"), fine meshed strainer, some dish washing soap and an old, stiff toothbrush. Begin by placing the strainer over the measuring cup. Pour the contents of the glass jar through the stainer into the measuring cup. Be absolutely sure that all of the tiny metal parts such as the hookkeeper strap and ring come out of the jar and catch in the strainer.

When you are satisfied that all of your hardware is in the strainer, pour the lacquer thinner from the measuring cup back into the glass jar and close the lid. That should help that headache you are beginning to feel. Did I mention that you should do this in a well ventilated area? Now I have. Go to a sink and run the water until it is very hot. Add a squirt of the dish soap to the contents of the strainer, and hold under the stream of hot water. This will help remove the old varnish and gets rid of the lacquer thinner that was left behind. Now dump the contents of the strainer onto an old towel. Use the toothbrush and a little more soap to brush off each guide carefully. *Don't drop them into the drain.* When each is cleaned and rinsed, place on a paper towel to air dry thoroughly. Sometimes you may want to use a bit of 0000 steel wool to clean the guides a little more, especially the stripping guide. Pointed ends of a round wooden toothpick are also useful for removing stubborn varnish from guides.

If it is an agate guide, be sure all varnish residue was removed from the inside diameter. Steel wool won't hurt the agate if you are very gentle. Agate guides are a bit fragile, so don't drop them onto hard floors and table tops. Never fish with a cracked agate guide, it will chew up your fly line. Of course, a cracked agate guide is okay for a wallhanger.

Sometimes stripping guides and tip-tops were oxidized. If so, you can use the same methods described for ferrules to darken them if necessary. If the guides have rusted, you may want to scour off the rust with steel wool, but guides that have been rusted should never be re-installed on a rod intended for fishing. Consider them as last choice to be used for wallhangers only. It will maintain historic accuracy and interest of the piece.

Wrapping and Spacing Guides

How many guides are enough? How do I know where to put them? How big (or small) should they be? What kind of thread do I use? How do I make the color stay bright? How do they make the thread wraps look so tight and neat?

Have you asked these questions? Most of us have wondered about them from time to time. I'll let you in on a secret; it ain't rocket science. Somehow, many of the procedures involved in finishing a rod seem to have acquired the status of *trade secrets*, and are guarded more closely than the crown jewels. It is difficult to imagine just how this came about, but it seems that the general feeling is that, *If folks knew how easy it can be, everyone would be doing it.* That's not quite true – for many people, it is time, more precisely the lack of time – that will prevent them from ever trying their hand at rod restorations.

Let's begin with the number, size and spacing of guides. The major concern of most rodmakers in deciding how many guides to use was the intended selling price of the rod. No, the guides aren't that expensive, but the time required to wrap the rod is increased by each added guide. Simply put, it takes more time to wrap a nine-foot rod with ten guides than it does to wrap a nine-foot rod with six guides. Math tells us it could take 40% less time for the lower number. And as labor became more and more expensive, so did the cost of wrapping each guide. Note that as costs rose, the first thing to be eliminated by rodmakers were the classically beautiful yet time-consuming intermediate wraps. Some early Leonards, Paynes, Kosmics and other high quality rods had as many as 150 of these tiny wraps. As late as the mid 1920's, one model of Divine Rods - the Gloriwest - had 275 intermediates, not counting the extra tip!

They were necessary in the early days of rodmaking. Animal-based glues were unreliable, and these closely spaced wraps had the function of helping to hold the six strips of cane together. The appearance must have pleased fishermen, though, as even wood rods (made of a single rounded shaft of wood) were wrapped with intermediates for ornamental purposes. These wraps were replaced by tipping and a butt wrap pattern that accomplished the goal of making the rod look *fancy* without taking all day to accomplish. If intermediates were used at all after about 1930, it was usually on a higher grade rod, and the wraps were more widely spaced.

Time is money, so rods destined to sell at lower prices were equipped with fewer guides than their higher-priced counterparts. In many cases, notably rods by Goodwin Granger, the *only* difference of consequence to the fishing characteristics of the rod was the number of guides used. Other minor differences were simply cosmetic, and made no difference. In the Heddon line of the 1930's, there was a Model 21 Featherweight rod that sold for ten dollars less than the Model 31 Featherweight simply because the #21 was not in a case, just a cloth bag.

Sure didn't make the rod fish any different, did it? Moral of the story: rods that cost less did not necessarily have less desirable qualities. Other ways used by makers to lower costs included use of cheaper grades of cork for grips, NPB instead of N/S hardware, less attention to node spacing and the packaging of the rod. But, we are concerned here with the number of guides used and how to space them on a rod.

If you are restoring a rod, it is a simple matter of replacing the guides in the same locations where they were originally placed. Use the map of the rod you made before you started. If you are returning a rod to its original condition, you can probably find the original location of the guides in one of two ways. First, look on the rod shaft for dark spots left by the feet of the guides. If the ferrules have witness marks, you can use them to be sure you are looking on the proper flat of the rod. In most cases, some kind of mark was left by the guide feet in the form of an oxidized metal *stain* on the wood. Steel guides popular until the advent of Stainless Steel made a nice mark. Later guides did not leave as much of a stain, but very often the wraps of the original rod left a light spot on the cane. This faint impression is called *ghosting*, and is visible on most stripped rods. In the case of some very old rods, the thread may have even left a colored stain where it was wrapped and color preserved. This seems especially true of rods made of Calcutta Cane. Perhaps it was due to the nature of the cane; maybe the thread dyes from that era, or even the type of preserver or varnish that was used. No one seems to know. Whatever the cause, these are helpful in determining the original wrap patterns. Sometimes you can even see the signature wrap pattern on the butt, and the location of intermediate wraps.

If you are stripping a rod for restoration, ***Do not count on these marks showing!*** If you do, they will not be visible. Trust me on this one. *Always* map each rod that you strip from original condition! On the other hand, if you are not refinishing a rod to original, these marks will show up under the new varnish and advertise that it is not wrapped as original. Keep that in mind when making your decisions regarding how to redo the rod.

Another way to tell where the guides were located is to look for mars in the cane where someone previously removed the guides. Odds are that if the previous refinisher did not redo the rod as original, he also wasn't very careful in removing the guides and may have left razor blade cut marks where the guides were removed. This occurs all too frequently with amateur refinishers.

If you are refinishing a rod of a lower grade and wish to upgrade it for fishing by adding additional guides, the best bet is to copy the spacing of the guides on a rod that casts well and is of the same length. Place the guides on the rod shaft temporarily with small rubber bands. Make the rubber bands by cutting a length of ¼" rubber surgical tubing into bands about 1/8" wide. These small rubber bands will allow you to cast the rod and move the guides to the location that seems to give the best performance. Put a weight on the end of the fly line and put a typical arch in the rod to see that the line does not contact the rod at a *flat-spot*. If it does, move the guides, or add more, so that the line is held away from the rod. Sounds complicated, but when you try it, you'll see how easy this can be.

The rule of thumb for the number of guides a rod should have is that there should be one for each foot or part of a foot in length, not counting the stripping guide. Thus, an eight-foot rod should probably have eight guides, and an 8½' rod should have nine guides – five on the tip, four on the mid, the stripper only on the butt. When planning the number of guides to buy, don't forget to buy guides for the extra tip.

Some rods will seem just fine with fewer guides, some may even seem to need more. Experiment until you find what works best. When you are finished, map the rod for guide locations, this will save you time on the next rod!

Now, on to the style and size of guides to use. Modern guides are not available in the same styles and in some of the sizes as the oldies. If you are replacing the guides completely, this is no problem. If you are attempting to restore a rod, it can be a crisis. Snake guides are sized by a comparative number. The largest guides are the whole numbers, beginning with #6, then growing smaller: #5, 4, 3, 2, 1, and then 1/0, 2/0, 3/0, 4/0. Most classic rods used guides of sizes down to 4/0. These are tiny, and were great for use with the small diameter braided silk lines of the early 1900s. They are not as good for use with today's tapered lines, especially the weight-forward types that are so popular. The manufacturers who did specify a line size usually recommended a DT line by the old letter designation, such as HEH, or a level line, such as E.

Until you build up a supply of *spare parts*, you may find it is better to replace these guides with the closest available modern snake guides. English style (reverse-twist) snake guides are not currently available from any modern supplier, so you may have to compromise the style as well. Stainless Steel guides by Perfection, the industry standard for years and years, are still available in sizes down to 3/0, but can be difficult to locate. Contact Perfection Tip Company for suggestions or to order in quantity. Most suppliers can provide you with Perfection snakes from sizes 2/0 and up in either chrome or black finish. Snake guides from Pacific Bay are a good substitute for Perfection. Tip-tops have not changed much for the last 75 years, so they are easy to replace. Just be sure that you use the small loop tops, as they are in the style of the classics.

Rule of thumb for snake guide sizes: If the stripping guide is a snake guide, it was usually a #2. Use no larger than a #1 for other butt and mid guides. Usually a 1/0 is proper. For tips, graduate the size downward as you get closer to the tip, beginning with 1/0 and going down to a 2/0 or 3/0 if available.

Stripping guides for classic rods tended to be smaller than today's guides as well. Most were a Perfection guide in chrome, steel, Stainless Steel, or if black, were carbide. The most common sizes used were 6mm and 8mm. These represent the inside dimensions of the guide opening, and are still the size designations used by manufacturers. Today, the smallest available is usually an 8mm, and is found under the section of catalogs devoted to casting guides. Perfection was the most common guide, but Mildrum was a popular manufacturer of stripping guides, and is still producing a similar product. Be sure to order an *unbraced* guide.

Agate stripping guides are scarce at this time, though there is reported to be a new supplier coming on line. These classy guides can be salvaged from old butt sections. Agate stripping guides are fine as long as they are not too small for the fly line to easily pass through the opening, but agate tip-tops are not recommended. They tend to be too small, and are easily broken. Be careful with all agate guides. If they crack (and they are brittle and delicate) they will tear up a new fly line in just a few casts. A short discussion of agate style guides is in order; they were not all created equal. The best agate guides were made of genuine agate. They were mounted in fine frames of nickel silver wire. The best of these had a substantial, sometimes rounded, frame around the agate itself which helped to protect the agate as well as adding to the elegance of the guide. Look at the original agate stripping guides on rods by Leonard, Payne and Goodwin Granger for examples of the best quality. Lesser agate guides had a simple wire frame that fit into a groove in the agate ring. These were used on medium quality rods. Lower quality agate guides were mounted in a stamped frame of N/S or other materials. These frames are flat, as are the guide feet, and are usually seen on cheaper rods and on bait and casting rods.

As stated previously – all agate guides were not created equal, and some weren't even agate! The best clue that a guide is not agate is the color. Genuine agate was used in many colors. The color may be white (nearly clear), orange, amber, red-orange or deep red, like Burgundy wine. Agate has a *variable transparency;* that is, it is darker in some spots than others and looks vaguely milky when you look through it at a light source. *Fake* agate is really glass. It is too perfect; the color is usually bright ruby red with a consistent transparency and no milky quality when held up to a light. Some makers of these guides were sneaky and the *agate* was treated to make it look grayish around the edges, until held up to a light. While you are holding the guide up to a light, be sure to look for cracks. Do not use a cracked agate guide on a rod that will be used, but it's okay for a wall-hanger. This fake agate isn't bad. It just is not genuine, and is less classy in every way. But it is still prettier than a plain old Perfection stripper!

Simulated agate was given many names by the rod company copywriters: Agatine, Crystal Agate, and Genuine Garnix, were among the common names. Read rod company catalogs carefully and you'll see that the higher grade rod's descriptions were careful to state *Genuine Agate*, it was a mark of pride, craftsmanship and quality.

Wrapping of guides is just like anything else - you need the right tools, the right procedure and a lot of practice. For procedure, read the books by Cattanach, Kreider, Garrison & Carmichael and Kirkfield. They are thorough, and each has a few tips that will smooth the learning process. When it comes to the proper material for wrapping thread, there is no question that pure silk thread is the best as well as the traditional choice.

There is a fine manufacturer of silk thread from Japan that has a distributor in this country. The distributor is the YLI Thread Corporation of Provo, Utah. YLI can supply you with 95 percent of all the colors and sizes of thread that you will need. They have a premium quality product that is reasonably priced and will provide you with good, prompt service.

There are three traditional sizes of thread: the largest thread size you should use is size "A", which is the standard sewing size and the most commonly seen thread. You will use this size most of the time. The next most common, and usually the best size for fly rods, is a smaller diameter, called "00" or 2/0.

Least often used, because it can be hard to find, but great for tiny intermediate and tipping wraps, is size "6/0". There is only one supplier at this time, Pocono Gateway. The 6/0 thread is made by Pearsall's, and is called *Gossamer* for good reason. This thread is so fine it makes an 8x tippet look like an anchor chain. It is especially necessary for the elaborate and complex signature wraps on Leonard Catskill rods of the period 1900 to 1920's.

Thread from YLI comes in two diameters, listed by European designations. Size 50 is just slightly smaller than "A" thread, and is available in nearly every color you ever need. Size 100 is roughly the equivalent of 000 or 3/0 and is a great thread for tipping and intermediate wraps, although the larger size will be fine, too. Some authors have made the use of silk thread sound difficult, but it is actually easier to wrap than nylon as it will lay flat and will wrap closer than modern materials.

If you are completely re-wrapping a rod, choose the color that best duplicates the original thread color from the charts. If you are simply replacing one or two wraps, you'll want to match the color of the thread ***as it now appears on the rod***. It matters little what the original color really was; try to find a color that, after color preserving and varnishing, will match the original wraps on the rod. If you are unable to locate a color that is an exact match, use the closest shade available. When the guide has been wrapped and before you color preserve it, dye the wrap with water based marking pens. Usually the best marker colors for matching old wraps are shades of yellow, as this approaches the color of ambered varnish. Sometime you may want to use gray markers to darken the thread slightly. After using the marker, ***you must wait for the ink to dry completely*** before you will know how close your color match really is. Remember that wet thread is much darker! When you are satisfied with the color match, you may proceed as usual with sealing the wrap.

Some later cane rods were wrapped with nylon; most Phillipson rods for instance. If that was the original material, you should use it. With the sources listed herein, there is little excuse for not using the right type of thread! If for any reason you cannot get the right silk color, it is better to use nylon than to use regular sewing thread of Mercerized Cotton or any other modern material. This is because nylon is not as *fuzzy* as these other materials. It is difficult to tell nylon from silk after the wraps are color preserved and varnished, but that isn't the point. It is right to use silk in restoration, and it is not a restoration if you don't.

To begin wrapping the rod, mark the guide locations on the proper flat of the rodshaft with a pencil. You should hold the guide in place temporarily by taping one foot to the shaft with Scotch or masking tape. You should tape on only one guide at a time - the one you are about to wrap - and after wrapping the first guide foot, remove the tape and wrap the other foot. When you wrap a rod, it is best to wrap the guides in place first, then later you can add the intermediates and ornamental wraps. If the guides have tipping wraps, add these immediately after wrapping all of the guides, then add the color preserver to the tipped wraps as a unit. Sealing and color preserving the wraps is not a big project.

There are three basic treatments and each one leaves the thread looking different on the finished rod.

First and most common on classic rods was lacquer color preserver. Lacquer makes the thread look almost the same color as it appears on the spool. It will make a bright and colorful and opaque finished wrap; the appearance favored by most of the classic rodmakers. When preserved with lacquer, silk thread has a sparkle and *snappy* look that is very attractive. The best source of the right kind of lacquer is a genuine paint store. Use a lacquer that is intended for brushing rather than spraying; you'll have better results. The lacquer should have two traits - it should be absolutely clear, even when dry, and it should be somewhat flexible. You will need this lacquer in three thicknesses and you can mix them yourself in small amounts. The first coat should be mixed half-and-half with the proper lacquer thinner. The next coat is one-fourth lacquer thinner and three-fourths lacquer. The final coat should be straight lacquer. But as you use it, lacquer in the bottle will thicken, so pay attention to its original viscosity and try to keep it at that thickness. When lacquer gets too thick it is very difficult to get good results.

After you finish wrapping the rod, burnish the wraps with a smooth, rounded tool. The finger hook on scissors is good for this. The object is to smooth the thread, make it lay flat, and push the individual wraps closer to each other. Don't get carried away or you will actually separate the wraps and make the thread fuzzy. After burnishing, singe the thread by holding it over the flame of the alcohol lamp for a couple of seconds. If you rotate the rod rapidly and don't leave the thread over the flame too long, any loose filaments will burn off and the heat will shrink the thread and make it tight to the rod. If you heat the thread or the guide too much, the thread will burn, so be quick and careful. You will notice that the guide feet are clearly outlined under the tight wraps.

Singe all of the wraps on a rod joint then *immediately* give the wraps a coat of the thinnest lacquer solution. This will prevent more filaments from fuzzing and will further shrink the thread as well as binding it to the rodshaft. When you start to wrap intermediates, do a few (whatever goes between two guides), singe them and lacquer them quickly so that they will not *blow out* and create more work.

With any color preserver it is ***very important not to let the preserver get on the cane.*** Apply it with the Cat's Tongue brush or the Spotter to give you better control. The thinnest lacquer is used first to accomplish two things; one is to give complete penetration into the thread and the other is so that it will dry within a minute or so to allow you to keep working. When all of the wraps are complete on a rod section, use the next thickest lacquer to give the wraps another coat and let it dry for 24 hours in a warm place. It is not necessary, or even desirable, to put on a heavy coat. Just be sure that you apply a thorough coat. You can add as many coats as you want if you desire a *high-build* finish on the wraps, but the classics did not have a thick coating.

After the first two coats have dried, use one coat of the full strength lacquer and let it dry completely. Until the rod is varnished, you should refrain from flexing the rod as this will make tiny cracks in the lacquer and will create spots for varnish to seep under the lacquer and spoil the appearance of the wraps.

The final treatment of the wraps is a light coat of varnish. The varnish should be applied in a warm room - at least 72°. Brush a coat of spar varnish on the lacquered wraps, not too heavy, and set the rod aside to dry for two days in a warm and *dust free* location. The varnish should dry to the point of being dust-safe in a matter of five or six hours, but do not handle the varnished areas or let them touch anything for at least two days, and a week is even better.

The second type of preserver used by the classic rodmakers was clear shellac. The shellac preserver gives the thread an entirely different look. Silk preserved with shellac is semi-transparent, and a bit duller in appearance; not quite as gaudy and bright. Shellac must be thinned in the same manner as the lacquer for good penetration, and takes longer to dry. Otherwise, use the same procedures. Be sure the final finish you plan to use is compatible with shellac; many are not. Shellac had the advantage of not being as volatile as lacquer, Remember the famous fires at factories such as Leonard, Phillipson and Divine?

The third treatment is the easiest, cheapest method, and the most forgiving. This method uses the same finish that will be used for the final coat. For example, if you plan to use spar varnish, use the same exact varnish for thread sealer. If you use polyurethane, use it as a sealer. It is best not to mix the two as they do not always bind well to each other, and in some cases will not bind at all.

This method is *forgiving* because the wraps need not be perfect to look great when finished. The result of using a final finish as a thread sealer is interesting and beautiful. It leaves the silk thread looking like a transparent jewel coating the guide feet. The only possible drawback cosmetically is that the guide feet show through the thread quite clearly. Some folks find that objectionable, others like it. Again, if you are restoring a rod, do it as original. If it's a fishing rod, experiment and have some fun. Red, scarlet, gold, yellow, emerald and some others are stunning on light cane; tan, gold, goldenrod, red, primrose, copper and orange create a rich look on darker or flame treated canes.

There were few classic rodmakers who used this method. The Fred D. Divine Co. and Thomas Chubb used this process on the entire rodshaft for the Silkien Process rods (Patented in 1897 by John Kenyon of Toledo, Ohio). Every rodmaker uses the process for *invisible wraps* over damaged rod sections. The only maker who advertised *clear guide wraps* was, once again, the Fred D. Divine Co. on the 2-piece Fly Wate rods of the late 1920's. Although E .E. Garrison and Lyle Dickerson used the same method, they did it for a reason other than the cosmetic appeal, the varnish virtually welded the thread to the cane.

To accomplish these clear or transparent wraps, simply proceed as usual, but the first coat of sealer should be carefully worked into the thread. It is not a problem if this gets on the cane, just wipe it off as best as possible. Work the finish into the thread with your finger to assure good penetration. Let this first coat dry for a few days before applying a second coat *or* follow the re-coat instructions for the specific finish you are using. Usually two coats of the finish are sufficient before applying the final finish to the entire rod.

If you are using this method for invisible repair wraps, here's another tip; use white thread on light colored rods, and for darker rods such as Heddon, Granger and Phillipson use an ivory to very light tan. Also, do not wrap the thread under too much pressure or the varnish will not penetrate thoroughly and you will end up with a hazy look to the otherwise invisible wrap. Usually, the wrap will not look entirely transparent until it has about two coats of finish in place, so don't worry if it still looks opaque after the first coat dries.

Some method of keeping a steady pressure on the thread coming off of the spool must be used to get good results. Some makers simply put the thread in a coffee cup and strung it through a phone directory to apply pressure. A sewing machine type thread tensioner is also a handy tool, and is available through rodmaker's supply houses. Orvis sells a thread tensioner mounted on a C-clamp. It works well and is easy to learn to use.

One final tip: some days you'll be able to wrap quickly and accurately with great results, but some other days, you are all thumbs. Find out which kind of day it is going to be, and only wrap on "good" days! You will save a lot of silk and your sanity if you follow this advice.

Varnishing Bamboo Rods

Perhaps more than any other phase of restoration, varnishing the rod is intimidating. It shouldn't be. Rodmakers and restorers have for many years given the impression that you must build and use some sort of elaborate high-tech contraption in order to create a good varnish job. It just isn't so. This job is only as difficult as you make it.

It seems that the complexity of varnishing rods grew with the promulgation of Garrison and Carmichael's book. Garrison's dipping mechanism was quite effective, and being an engineer, it probably made sense to him. Most of us don't need to get so involved; what do you think they did in the 1870's and 1880's to get a good varnish job? You can bet that they simply used good woodworking skills and practices to accomplish their task, and so can you.

Varnishing by any method has a set of rules that ***must*** be followed. The same basic rules always apply. Begin with a dust free and warm room with no drafts. Best results with varnish seem to be had when the room is at about 80°F and the humidity is 30-50%. The varnish should be at, or above, room temperature. To warm the varnish (other than a spray can), place it in a pan of hot water on a warming device. Stir the varnish before use. Usually, if the varnish is stored in the room where the varnishing is to be performed, it will be warm enough.

Before you begin, wipe down the entire rod with a tack cloth. Then wipe with a soft, lint-free cloth dampened with denatured alcohol. This will remove any grease spots left by handling the rod. Don't soak the rod, especially the varnished wraps. The alcohol will damage the varnish if you use too much or rub too hard. Just lightly wipe the entire rod, and then don't handle the areas to be varnished.

After varnishing, it is helpful to have some sort of cabinet or empty closet where the varnished rod may be left undisturbed for at least one week. Varnish will not fully cure for several weeks, or perhaps even several months if the humidity is high and the varnish is heavily applied. After one or two weeks, though, normal handling is okay, and the rod won't attract and hold dust. Always follow these rules and you will be off to a good start.

Several methods may be used to apply varnish. Also, there are two schools of thought regarding when to varnish; before or after wrapping on the guides. Some prefer varnishing the naked rod blank. There are a couple of advantages. If the job is terribly botched or the varnish crawls or wrinkles, you can simply strip the rod and start again. It is also easier to polish out the varnish if you have no guides and wraps to work around. The main disadvantage to this method is that it can be difficult to wrap the guides on over the varnish. It is up to you to try both ways and see which you prefer.

Actually applying the varnish can be accomplished in four different ways: dipping, dripping, spraying or brushing. The classic makers probably used brushing more often than any other method. Some of them were magicians with a brush, and created flawless, glass-like varnish jobs. See the book by Garrison & Carmichael for some good tips on brushing varnish. Did they polish the final varnish? Who knows? Anyone who does know won't tell. Try it - maybe it will work for you.

Many production rod companies used the spraying method. This made sense for outfits that were turning out hundreds of rods, but it also makes sense for someone who will only be varnishing rods on an occasional basis. This can be very effective, and it is a cost-efficient way to get the job done. It will work for you if you do it right. What is required? You will need to experiment with the can of spray varnish. Perform your tests on a dowel - it's like a rod for test purposes, but a lot cheaper. What you are testing is the distance to hold the can from the rod to get a nice, smooth coat without getting an *orange peel* effect. If you've never spray painted before, try this: Intentionally hold the can too close to the object to be painted and spray. You will note that the finish has a speckled effect that looks just like the texture of an orange peel, hence the name. This is caused by the propellant in the spray can *denting* the surface of the finish. It can be remedied by holding the can a little farther from the object to be sprayed.

It is sort of a game, though, because it is better (for purposes of control) to be closer to the rod; so you have to balance the two. Usually, the instructions on the can will say something like "hold the can 11 or 12 inches from the surface to be painted." Be sure to follow all instructions suggested by the manufacturer of the spray varnish. Before you begin, mask off ferrules and grips with masking tape. Be sure that the opening of the female ferrule is completely sealed! Wrap the grip with paper and tape the paper in place. You can tape a paper clip to one end of each section so that you can hang the rod to dry. Use a large cardboard box as a spraying booth (or backstop) to catch the overspray. Spray on a very light first coat, let it dry and be sure to re-coat within the manufacturer's suggested time. Usually two coats will be sufficient.

After the second coat, let it dry for at least a week before handling. If there are any little rough spots, use the Brownell's polish to remove them. (See the chapter on Polishing Rods.)

Dripping on the varnish is a seldom-used but cost effective way to get good results. It is usually done before wrapping on the guides. Suspend the rod for this procedure: for the butt section, ferrule down; for the tips, ferrule up. Remember to mask off the ferrules, especially the females. Place a *clean* coffee can below the first section. Using a turkey baster, drip the varnish onto the rod from the top and let the excess flow into the can below. It is best to have very warm varnish for this procedure. Some recommend thinning the varnish with one part thinner to four parts varnish so it will flow better. As a rule, varnish should be used as it comes from the can, but in this case it seems to do better if thinned.

When the first section is thoroughly coated, move the catch-can to the next section and continue until all sections are coated. Let the varnish dry adequately, at least a week before handling.

Perhaps the most effective and foolproof method for varnishing is to dip the entire rod, with the guides already wrapped in place. The bad part is that if there is a major problem, you may have to repeat all of the work over again. Just be sure to do everything right up to this point, and you should have no trouble. There are a lot of complex contraptions recommended for this procedure. They are fine, and work well, but they are not necessary to do a very respectable job.

The simplest of the dipping systems is detailed in Kirkfields's book. Borrow a copy if necessary. Kirkfield is absolutely correct when he states that the speed of withdrawal from the dipping tube is the most critical part of the job. It is. Experiments have shown that it is not the exact speed that is important so much as having a consistent, steady withdrawal speed.

Motors designed for rotating graphite rods while the finish dries are easily obtainable from the rod building supply houses. Another good motor is the one used on barbecue grills to turn the food over the coals. Simple mechanics can be used to increase or decrease the speed of withdrawal. As the rod is withdrawn, blow the varnish out of the guides, or use a rolled paper towel to carefully blot the varnish away from the guides; the details are in most rod building books. It is quite simple to do a fine job with some practice. Use dowels as *rods* for practice. You can make your dipping apparatus as simple or as complex as you wish, but the most important areas to control are temperature, humidity and speed of withdrawal. Then practice, practice, practice.

The type of finish to use is often a topic of dispute. For originality, spar varnish is the usual choice since 99 percent of the classic rodmakers used this finish.

Its advantages are that it dries to a tough, flexible finish that is easily polished to a glass-like gloss. The disadvantages are that it is touchy about the temperature at which it is stored, as well as humidity. Additionally, spar varnish takes a long time to cure to a hard, solid coat; sometimes months are required.

Spar varnish also tends to turn yellow, then amber, as it ages. Often a rod that looks like it was made of caramel colored cane will appear in its true nearly white color when the old varnish is removed. Many times you'll see a rod described as having wraps of yellow and black Jasper that is really yellowed varnish over black and *white* Jasper. Which thread would you use? Depends whether you are replacing all of the wraps or just trying to match one new wrap to the old ones, doesn't it?

Other choices for a final finish on the rod include polyurethane and Tung oil. Polyurethane is a fairly modern finish that has some great advantages. It applies quickly and easily. It is readily available. It dries very quickly to a very hard coat. There are two disadvantages to polyurethane, though, that make it less than ideal. It is not authentic for most classic rods and it cannot be polished well. The final choice is a *hand-rubbed oil finish* such as Tung oil. A few makers used this finish, so it is appropriate for their rods when being restored. Tung Oil is best, but Tru-Oil by Birchwood-Casey is fine as well. Both are available through Brownell's. For rods that you plan to rebuild in a modern style for fishing, Tru-Oil is a very good and easy to use finish. Whatever method and finish you settle upon as being best for you, don't be afraid to continue to experiment with other methods. Who knows? You might discover the *ideal* method for applying varnish and become a hero!

Independent Handles

Did you ever wade into a fast-flowing river to reach a rising fish in a tough spot?

If you have, you may have reached a point where the water was just a bit deeper and faster than you thought, and the mossy rocks on the streambed were a bit slipperier than before. You know what I mean; you are a little afraid to continue, but going back doesn't seem that attractive either. If you've ever been in a spot like this, and wondered not only *How did I get here?* but also *How can I get to a more secure spot?* you'll understand where this discussion is coming from.

I recently wrote an article about changing older 9-foot, 3-piece rods into Little Wonders, which are short fishable rods made from only the mid and tips of a bamboo rod. This is still not a bad idea, and one that is gaining in popularity among amateur rodmakers. Earlier in this book I said that originality is worth saving wherever possible. Yet with the Little Wonders I was not concerned about saving the originality of the rod.

So, here I am – in a position where it's tough to go ahead and even tougher to go back. Yet there really isn't a conflict of philosophy. But by way of clarification, I'll add one more part to this ongoing saga.

I still believe that in the case of the cheaper, long production rods it is better to modify them for fishing use than to have them collect dust in the garage or to be used as tomato stakes – especially if you chose a rod with a butt section that is so far gone it cannot be saved. As stated earlier, if the object is to make a rod fishable by modern standards, originality isn't that much of an issue.

However, some of the fine older rods longer than nine feet are very valuable in a historic sense. It would not be wise to damage their originality by adding a new, permanent grip on the mid section. At the same time, they are not much fun to fish due to slower actions and the heavy weight of long rods. So if your objective is to have a nice, shorter rod to fish, but you would still like to preserve its originality there is an easy solution that has been around for more than a hundred years. The answer to this dilemma is the independent handle.

One of the earliest examples of the independent handle was from Henry Wells' 1855 book *Fly Rods and Fly Tackle*. The idea was not new even then. Wells states *I am aware that this construction is not altogether new; but it is uncommon, while its great merit should make it universal.* He goes on to say, *Other than discarding the (ferrule) dowel pin, I believe this to be the most valuable improvement which can be applied to the fly-rod as at present made.*

Some big names in rodmaking must have agreed. By 1894, H. L. Leonard was producing at least three models of independent handle rods, including his famous *Tourist* series. In 1906, E. F. Payne bought the rights to a patent for an independent handle that was also reversible. The patented handle was designed by Chancellor G. Levison of New York, N. Y.

Dr. George Holden Parker in his 1919 book *Idyll of the Split Bamboo*, gives very precise dimensions and instructions for such an independent handle. He says, *The author wishes to emphasize at once his hearty agreement with the plan of the late Henry P. Wells, of fitting to the rod an independent grasp, chiefly for its eminently practical value.*

In defending the extra weight involved from the use of an extra set of ferrules, Parker says, *it is weight in a good place; furthermore, it may be offset by use of the extremely light yet very serviceable fly-rod reels that are available today.* One of the advantages that both Wells and Parker saw in the use of the independent handle was that amateur rodmakers needed to construct only one hand grasp, and could then use it on all of their rods.

The most recent rod company to use the independent handle was the Phillipson Rod Co. The 8½' Peerless Dry Fly Special came with an independent grip for use with the mid and tip sections. Thus, the rod was an 8½' 6-weight rod or a 6'3" 5-weight Midge rod. These rods were produced as late as 1970.

Making an independent handle is not difficult. The first determination is what type of grip to use. Wells recommended a wood grip and reelseat turned from a single piece of wood, usually red cedar. While Holden also saw merit in the use of the wood grip, cork was much more the norm of his era, so he included plans for a grip constructed of cork rings. See the illustration for details.

Today it may be difficult to come up with a short piece of bamboo rod section to use as the core of the cork handgrasp. Use a 3/8" dowel cut to the proper length as the core of your independent handle. You could even install a modern reelseat if you desire, and quickly and easily create an independent handle. Just remember, the handle ferrule must be sized to receive the male ferrule on the mid section you plan to use.

By using an independent handle, you can make use of an old longer rod without harming its originality. Some of the softer actions will produce a delicate, light-action rod when used in this manner. A way to have your cake and eat it, too.

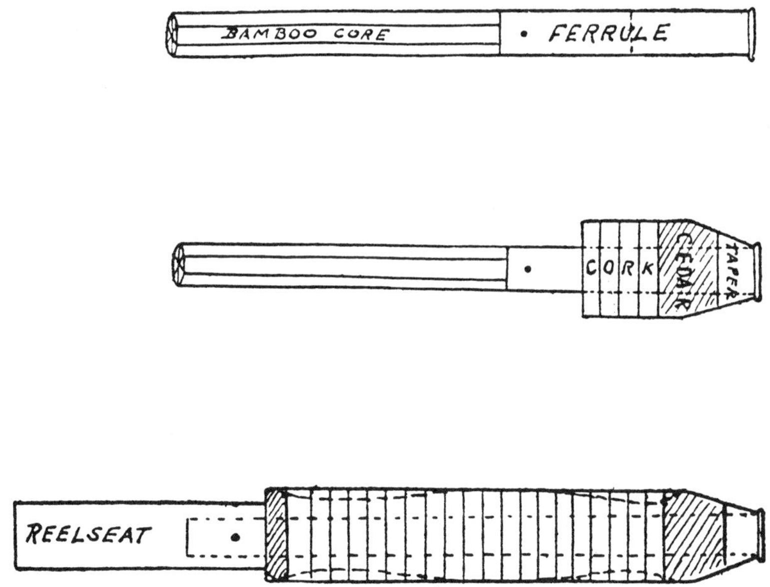

Vendor List

Silk Thread

YLI Thread Corporation
P.O. Box 109
Provo, Utah 84603-0109
1-800-854-1932

Britex
146 Geary Street
San Francisco, California 94108
(415) 392-2910

Pocono Gateway
1501 Interchange Road
Lehighton, Pennsylvania 18235
(610) 377-3441

Rod Components

Angler's Workshop
1350 Atlantic
Woodland, Washington 98674
(206) 225-9445

Custom Tackle Supply
2559 Highway 41-A South
Shelbyville, Tennessee 37160
(615) 684-6164

Dale Clemems
444 Schantz Spring Road
Allentown, Pennsylvania 18104
(610) 395-5119

Reelseats

REC Components
R.R. 1, Box 1605
Morrisville, Vermont 05661-9704
(802) 888-7200

Ferrules
(Super Swiss & Super Step-Down)

Bailey Woods
Classic Sporting Enterprises
Roaring Brook Road, R.R. #3
Barton, Vermont 05822
(802) 888-7859

Gunsmith Supplies
(Oxidizers, Polishes, Punches)

Brownell's
200 South Front Street
Montezuma, Iowa 50171
(515) 623-3896

Woodworker's Supplies
(Finishes, Lacquers, Shellacs, Moto-Tools)

Constantine's
2050 Eastchester Road
Bronx, New York 10461
1-800-223-8087

Nickle Silver
(Wire and Sheets)

Maroon Bells
3400 Tejon
Denver, Colorado 80211
(303) 433-2283

Tools - Router Bits, Etc.

Cascade Tools, Inc.
P.O. Box 3110
Bellingham, Washington 98227
1-800-235-0272

Machinery
(Lathes, Drill Presses, Sanders, Buffers)

Grizzly Imports, Inc.
P.O. Box 2069
Bellingham, Washington 98227
1-800-541-5537

Rod Tubes

Black Guard
3214 Matthew N.E.
Albuquerque, New Mexico 87107
(505) 884-7501

Project Rods, Bamboo Rod Blanks & Tips

Centennial Classic Sales
256 Nashua Court
Grand Junction, Colorado 81503
(303) 243-8780

Rods & Reels
17 Massasoit Road
Nashua, New Hampshire 03063
(603) 886-0411

Heritage Sporting Collectibles
22A Third Street
Turners Falls, Massachusetts 01376
(413) 863-3151

TRADE MARK
H U C B B
·T·H·CHUBB·ROD·CO·
MANUFACTURERS OF
·Fishing·Rods·and·Tackle·
"BUILT·ON·HONOR."
Post Mills Vt.

Montague City Rod Company.
MANUFACTURERS OF
FISHING RODS, REELS,
SNELLED HOOKS, LEADERS, ETC.

Guide to Original Production Rod Specifications

How To Use The Guide

This guide is useful in several ways. It will help you identify a rod that has lost its markings by comparing it to the drawings of rods, reelseats and ferrules and by the type of wraps. Grips usually have the measurements of the rings noted for accurate reproduction.

This guide can help you return a non-original rod to its original condition, including the proper wrap colors, style of guides and rodshaft markings. It can also help you determine the relative quality level of the rod to others by the same maker. That's where the original selling prices become helpful. Also, you may compare rods by different makers. A rod that cost $10 in the 1930s should be comparable to any $10 rod by any other maker in that era; intense competition made it necessary. Additionally, the information in this guide can help you determine the approximate date of any rod's manufacture.

The guide cannot show every rod by every maker, nor every variation of the hardware and wraps but it does provide reliable information for restoring a rod to specs that are accurate for the model at some point in its history. The information contained on the following pages comes from the catalogs of the manufacturers and sellers of rods as well as from the author's personal inspection of thousands of rods over the years. The major changes are shown in the charts, and the most common versions of the rod models are detailed.

This guide describes typical rods by each maker from each decade beginning in the 1920's and continuing through the 1950's. Some companies are covered in more detail than others simply because more information is available, or because the rods turn up with great regularity. Tips are provided to help estimate the date of manufacture by the style of hardware, wrapping colors and styles and rod markings. Where possible, changes in styles are noted by date in the text.

The short histories of the companies are just that – short! Enough information is supplied to help explain the different periods of the company (such as changes in ownership or management) and the reasons why changes may have been made. If you desire more information, refer to the bibliography for sources.

Perhaps one of the most useful charts are the ones called the *Flat Map*. This map is a graphic representation of the six flats of a rod. The front of the rod (the side where the reel is mounted and the guides are placed) is designated as #1, and the six flats are *unrolled* on the page to show the relative locations, style and information conveyed by rodshaft markings. For restoration purposes, the flat map provides the location of the hookkeeper (if there was one) and the decals. The decals are shown on the flat map in their approximate *life-size* and the proper location. Alongside the flat maps are enlarged versions of the decals used by the company.

These can be a great dating tool if they are present on the rods. The drawings are not perfect by any means, but they do show the wording seen on the decals and show the changes in styles through the years. The dates assigned to the decals are accurate to within a couple of years, as decals were often *phased in*. As with all components, the higher grades were switched to the new, improved style while the lower grades used up the supply of old components. The colors of the decals are discussed in the text for each maker.

Hardware specification sheets show the style of reelseat, grip, guides, tip-tops and ferrules. These are accompanied by full size drawings of reelseats that are referred to by name in the charts.

Wrap specification charts are of two types. The first is a graphic representation of the signature wrap or mid-butt wrap patterns. A *signature wrap* is located just above the grip, and may be fairly simple or quite complex. A *mid-butt wrap* is usually centered between the winding check and the stripping guide. You may lay your rod alongside the chart and copy the spacing, as these were created in that manner.

I have noticed that the spacings used by the companies for ornamental wraps seemed to change periodically, so if there are marks *ghosted* on the cane be sure to use them to accurately reproduce the original wraps. If you don't cover the ghosted marks, they will show on your restored rod and will detract from its appearance. Even if the pattern on the chart is not exact for your rod, the charts can still provide you with the proper thread colors.

The second type of wrap specification chart simply lists the model name or number and the colors of thread to use for the main wraps, tipping wraps, intermediates and some have notes on special wraps. If there is any conflict between the chart and the rod map you have drawn, you must decide which is better to use and which is likely to be the original pattern. Don't forget to take notes on any new wrap colors, patterns, rod maps or reelseat and grip styles you may encounter. You can add these to the information contained in this book.

Competition in the tackle trade made copywriters of catalog descriptions look for any and every way to make their own rods seem superior to others. For this reason, many *trade names* were created to dramatically describe a perfectly mundane item. Consider the common, inexpensive snake guide. Though most were made of Stainless Steel, various copywriters assigned the terms: Tungsten Steel, Tool-Hardened Steel, and Hardened Steel. They all designate the same quality, and were probably from the same manufacturer!

The same goes for treatment of metal work on ferrules and reelseats. Some makers referred to dark hardware as Oxidized, while others called it Blued, Gunmetal, Bronzed, or Blackened. Though different methods were used, they all produced the same result: black hardware that was less reflective than the so-called *bright* hardware. Bright hardware simply means that the metal was in it's natural silvery color, and was usually polished.

The same goes for wrap colors, as the copywriters used some fancy descriptions for silk colors as well. In the charts for each of the makers, I have elected to use the terminology that was used by them in their catalogs and advertising.

And finally, while I regret that this guide is not more thorough and complete, I am conceited enough to say that if someone had assembled one that was as complete, I would not have attempted it at all.

Abbreviations Used in Hardware Specifications

Types of Metal for Reelseats, Ferrules and Winding Checks:

Al = Aluminum
AnAl = Anodized Aluminum
(preceeded by color of anodizing)
NS = Nickel Silver
(German Silver, Liberty Silver)
CPB = Chrome-Plated Brass
NPB = Nickel-Plated Brass
BNS = Blued, Blackened or Oxidized NS
OBr = Oxidized (Bronzed or Blackened) Yellow Brass
BNPB = Bronzed Nickel-Plated Brass (Montague)
CPNS = Chrome-Plated NS
SCPNS = Satin-Chrome-Plated NS (Montague)
NS, GM = NS, Gun Metal finish

Materials for Strippers, Snake Guides and Tip-Tops:

SS = Stainless Steel
HS = Hardened Steel
THS = Tool-Hardened Steel
TS = Tungsten Steel
C or CP = Chrome-plated
B = Black colored metal
A = Genuine Agate
Ag = Agatine
G = Garnix
Cy = Crystal Agate (All Are Red Glass)

For guide abbreviations, the prefix :

P = Perfection Co.
M = Mildrum Co.
E = English or reverse twist snake guide
Am = American style snake guides
PC = Perfection chrome
PB = Perfection black
PSS= Perfection Stainless Steel
WS= Waldic Steel (really PB as above)
These are sometimes given to show the shape or style of the guide.

Reelseat Spacer Color Abbreviations:

B = Black	R = Red
BR = Brown	Y = Yellow
BRZ = Bronze	Wh = White
BU = Blue	M = Maroon
GR = Green	T = Tan
Gy = Gray	Co = Coral
N= Natural (bright aluminum)	

Reelseat Spacer Materials:
(May be followed by a color code)

HR = Hard Rubber	Bk = Bakelite
T= Tenite	P = Pyralin
Al = Aluminum	I = Ivoroid
Wal = Walnut	C = Red Cedar
CH = Cherry	Bt = Butternut
B Wal = Black Walnut	
C Wal = Circassian Walnut	
Comp = Composistion (Plastic)	

Ferrule Construction Methods:

RW = Rolled Welt
W = Welted, non-specified
HW = Hand Welted (Soldered Welt)
NW = No Welt
SH = Shouldered (bottle-shaped)
K = Knurled Pattern near base
KW = Knurled Welt
Hex = Hexagonal base, no ferrule station
WO = thread *Wrapped Over* ferrule
WP = Waterproof construction
S = Serrated
Mach = Machined Ferrule

Reelseat Styles:

SB = sliding band, usually all metal
C&R = cap & ring, spacer wood or metal
SDL = screw, down-locking
UL = up-locking
MLB = Montague patent locking band
LL = Lite-Lock (South Bend)
SL = Snap-Lock (Shakespeare)
PL = South Bend

Cane Colors:

N = Natural or honey-colored cane
BT = Browntone, caramel colored cane
BN = lighter browntone, heat treated
F = Random flamed cane
LB = Light Brown stained
MB = Medium Brown stained
DB = Dark Brown stained

Thread Color Chart

#	Color Name	Gudebrod	YLI		B – C	Britex
			#50	#100		
1	Black	001	Black	Black	BLK	B
2	White	002	White	White	WHT	W
3	Bright Yellow	8933	18	214		16
4	Honey Lemon	207	196	N/A	3010	17
5	Goldenrod	209	19	231		79
6	Gold	6038	79	N/A		200
7	Antique Gold		78	215		18
8	Orange	210	143	216	3155	N/A
9	Burnt Orange		N/A	N/A		165
10	Chinese Red	326	4	201	4350	33
11	Dark Red	595	3	N/A	4040	N/A
12	Garnet/Burgundy	5382	2	N/A	1063	97
13	Maroon	337	1	228		100
14	True Scarlet		5	202		Red
15	Primrose		169	217		158
16	Purple	468/474	22	244		182
17	Dark Blue	245	20	208		210
18	Light Blue	230	103	246		155
19	Lime Green	355	233	245	9124	178
20	Granger Green	356	764	236	9375	N/A
21	Dark Olive		156	247	9909	28
22	Light Olive		114	218	10332	9

Thread Color Chart

#	Color Name	Gudebrod	YLI		B – C	Britex
			#50	#100		
23	Emerald	577	30	222		95
24	Spring Green	358	N/A	220		21
25	Evergreen		112	221	9550	80
26	Copper		99	227	451	87
27	Beige	832	815	226		63
28	Medium Brown	541	83	235	5120	74
29	Chestnut Brown		100	234		92
30	Dark Brown	2619	820	N/A	5210	68
31	Gray	602	72	232	7820	54
32	Silver		70	224		41
X	Black & White Jasper		Jasper and Space Dye Threads Are Not Readily Available			
XX	Black & Orange Jasper					
XXX	Black & Yellow Jasper					
XXXX	Black & Red Jasper					
XXXXX	Black & Green Jasper					
Z	Rainbow Variegated Space Dye					
ZZ	Black & Yellow Space Dye					
ZZZ	Blue & White Space Dye					

Antique Rods

Antique Rods are so different in action and design from modern and classic bamboo rods that they are seldom used for fishing. Additionally, many of them are just too fragile to risk using them. If you have an antique that is sound and fishable, it can be very interesting and educational to try it out. You'll probably appreciate all of the advances that have been made in tackle more completely after trying out one of the old rods.

For the purposes of this book, a rod is considered to be an antique if it was made before World War I. Although a true antique should be more than 100 years in age, the author chose WW I (about 75 years ago) as the cut off date based on the number of changes that ocurred in the 1920's to modernize tackle.

Some of the makers of antique rods did not survive into the *Classic Era of Bamboo Rods*. These would include names such as Phillipe, Green, Murphy and Mitchell. Examples of rods by these men are rare and as such are seldom seen. If you come across any of these, treat them as historic treasures, and do not attempt any restoration. These must be inspected by rod historians for the information they can provide.

Other makers of the antique era did survive and flourish into the Classic Era. These would include names such as Leonard, Thomas, Payne, Divine, Edwards, Chubb, and Orvis. Of course, the earliest rods by these makers little resembled the classic rods they produced, though the overall level of their quality was present from the beginning. For example, early Payne rods were made by Edward Payne (Jim's father), and featured honey colored cane and usually had wraps of red tippd in black. They also tended to have all NS sliding band reelseats. This is quite different from appearance of the rods produced by Jim Payne in the Classic Era. Rumor has it that the earliest Orvis rods (like the one pictured on page 69) were made by Montague, though this has not been established with certainty.

The antique rods presented in this chapter were chosen because they show some of the typical features of the era before the turn of the century through WW I.

Most of the drawings were completed long before this book was planned, and for that reason, they were not of proper dimensions or format to be reproduced full size. The drawings are in their correct proportions, and where possible the percentage of actual size will be designated. Also, some of the rods will be presented in a format that requires you to turn the book 90° to view them and read the notes. This, too, is a result of using pre-existing drawings. My apologies for any inconvenience.

The general characteristics you will note on these rods are:

* Long reelseat, shorter grips
* Calcutta Cane
* Many have spiked ferrules
* Rolled and soldered ferrule construction
* Thinner cork rings in grips (Except the Orvis, which shows the earliest cork grip made of six strips of cork sheeting.)
* Very ornate decorative metalwork

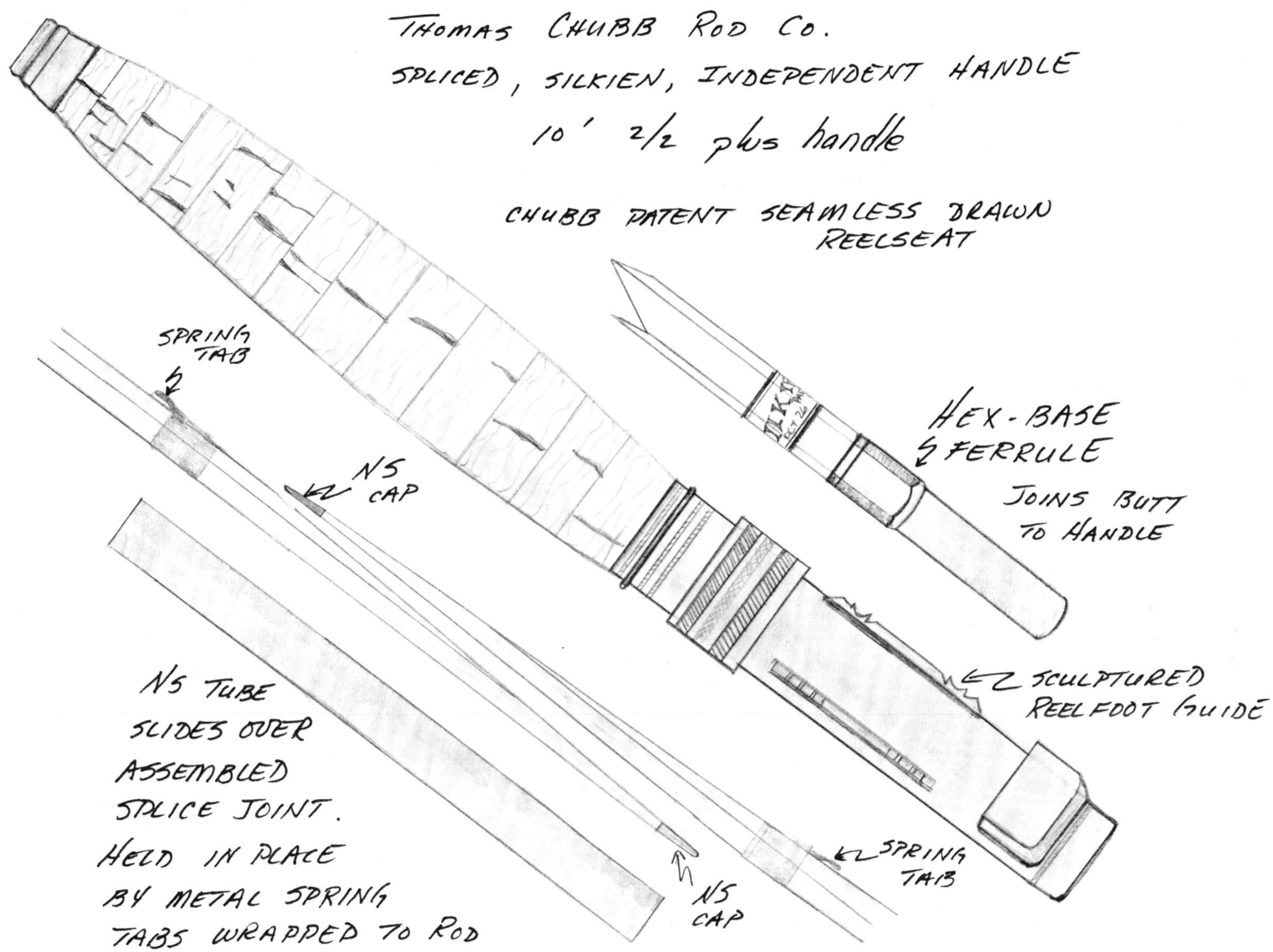
THOMAS CHUBB ROD CO.
SPLICED, SILKIEN, INDEPENDENT HANDLE
10' 2/2 plus handle
CHUBB PATENT SEAMLESS DRAWN REELSEAT
SPRING TAB
NS CAP
HEX-BASE FERRULE
JOINS BUTT TO HANDLE
SCULPTURED REELFOOT GUIDE
NS TUBE SLIDES OVER ASSEMBLED SPLICE JOINT.
HELD IN PLACE BY METAL SPRING TABS WRAPPED TO ROD
NS CAP
SPRING TAB

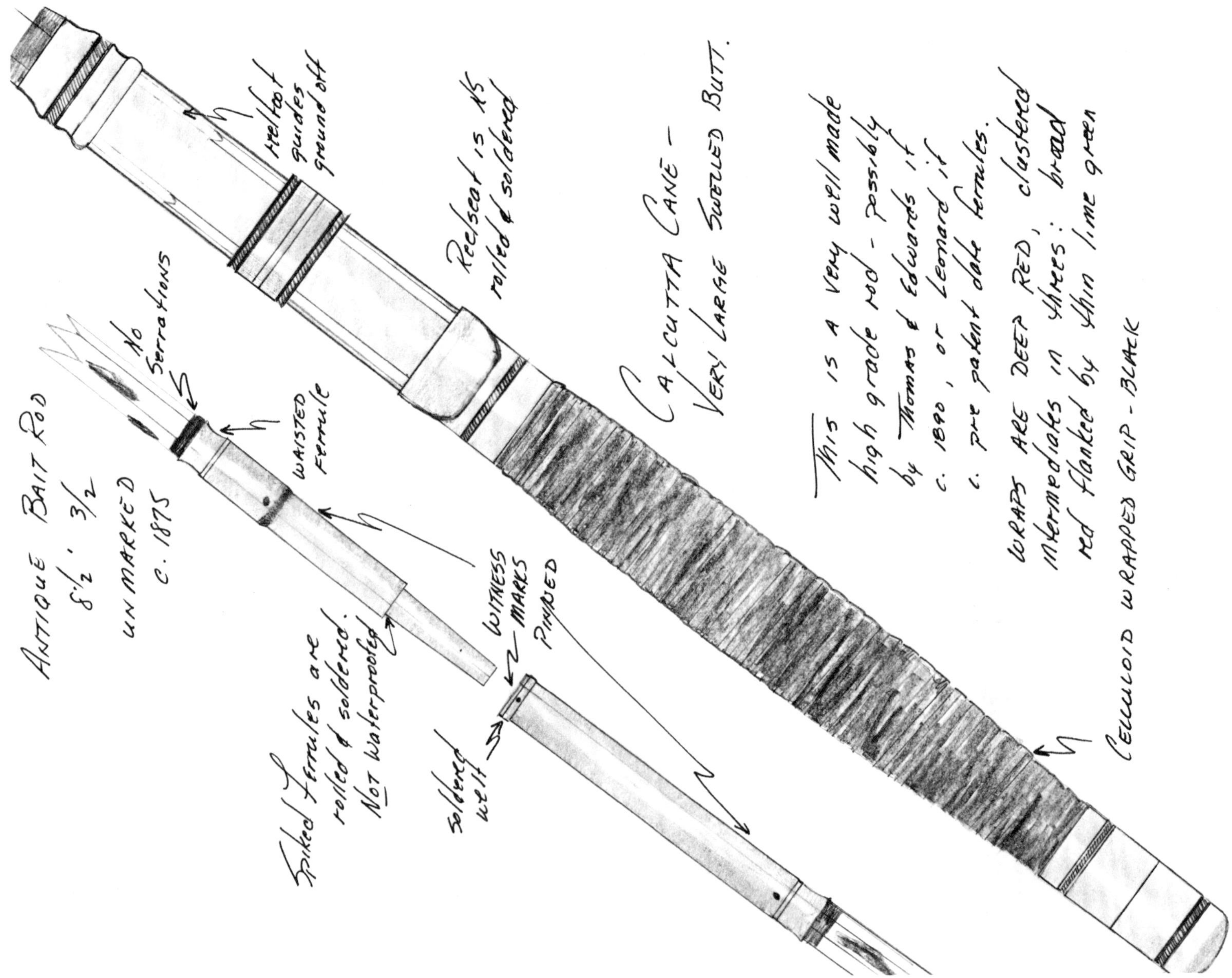

Antique Bait Rod
8½' 3/2
unmarked
c. 1875
No Serrations
Waisted Ferrule
Spiked Ferrules are rolled & soldered. Not waterproofed
Witness Marks
Pinned
soldered welt
steelfoot guides ground off
Reelseat is NS rolled & soldered
Calcutta Cane – Very Large Swelled Butt.
This is a very well made
high grade rod - possibly
by Thomas & Edwards if
c. 1890, or Leonard if
c. pre patent date ferrules.
Wraps are deep red, clustered
intermediates in threes; broad
red flanked by thin lime green
Celluloid wrapped grip - Black

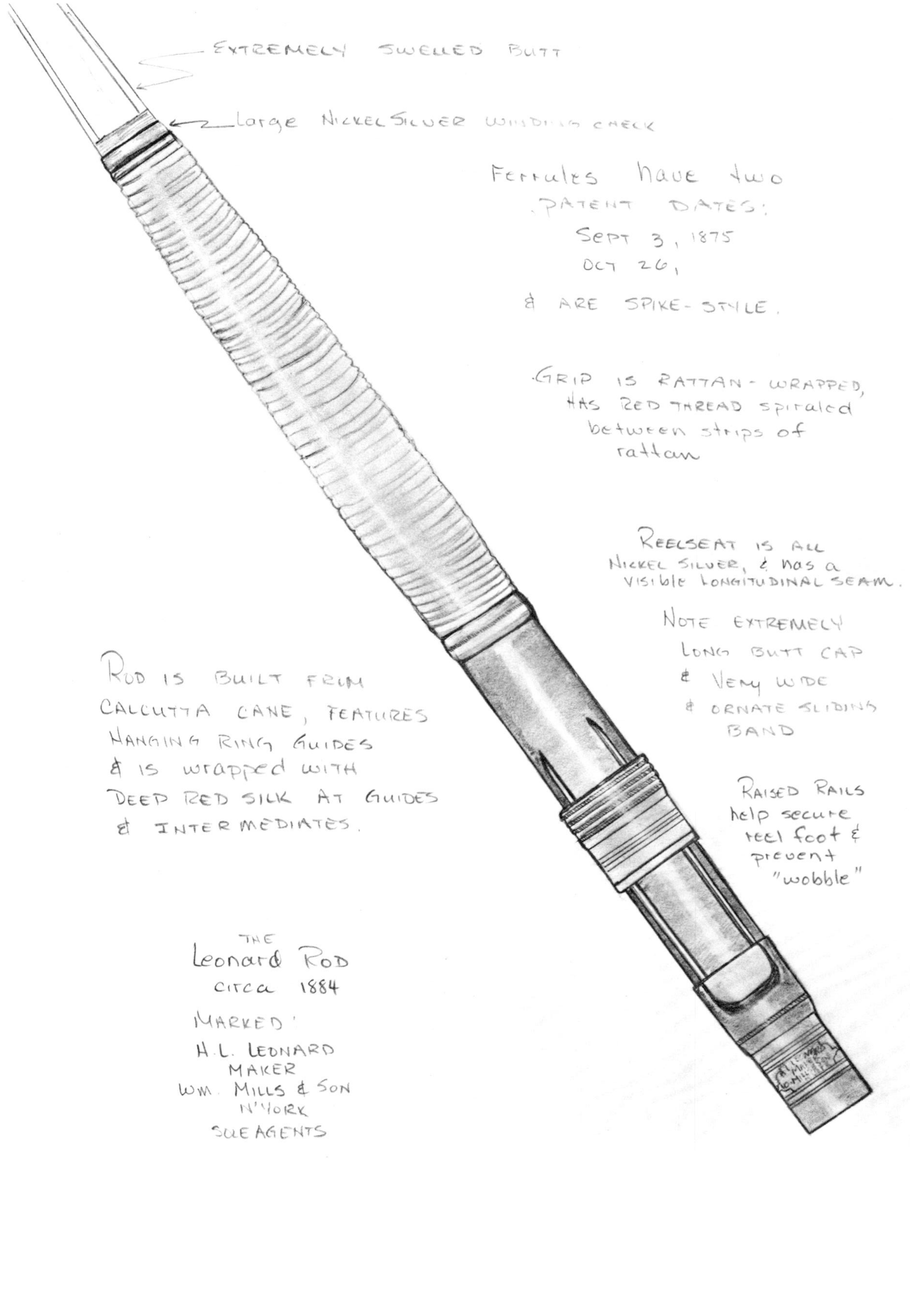
EXTREMELY SWELLED BUTT
Large NICKEL SILVER WINDING CHECK
Ferrules have two PATENT DATES: Sept 3, 1875 OCT 26, & ARE SPIKE-STYLE.
GRIP IS RATTAN-WRAPPED, HAS RED THREAD spiraled between strips of rattan
REELSEAT IS ALL NICKEL SILVER, & has a VISIBLE LONGITUDINAL SEAM.
NOTE EXTREMELY LONG BUTT CAP & VERY WIDE & ORNATE SLIDING BAND
RAISED RAILS help secure reel foot & prevent "wobble"
ROD IS BUILT FROM CALCUTTA CANE, FEATURES HANGING RING GUIDES & IS wrapped with DEEP RED SILK AT GUIDES & INTERMEDIATES.
THE Leonard ROD circa 1884
MARKED: H.L. LEONARD MAKER WM. MILLS & SON N'YORK SOLE AGENTS

H. H. KIFFE CO.
NEW YORK

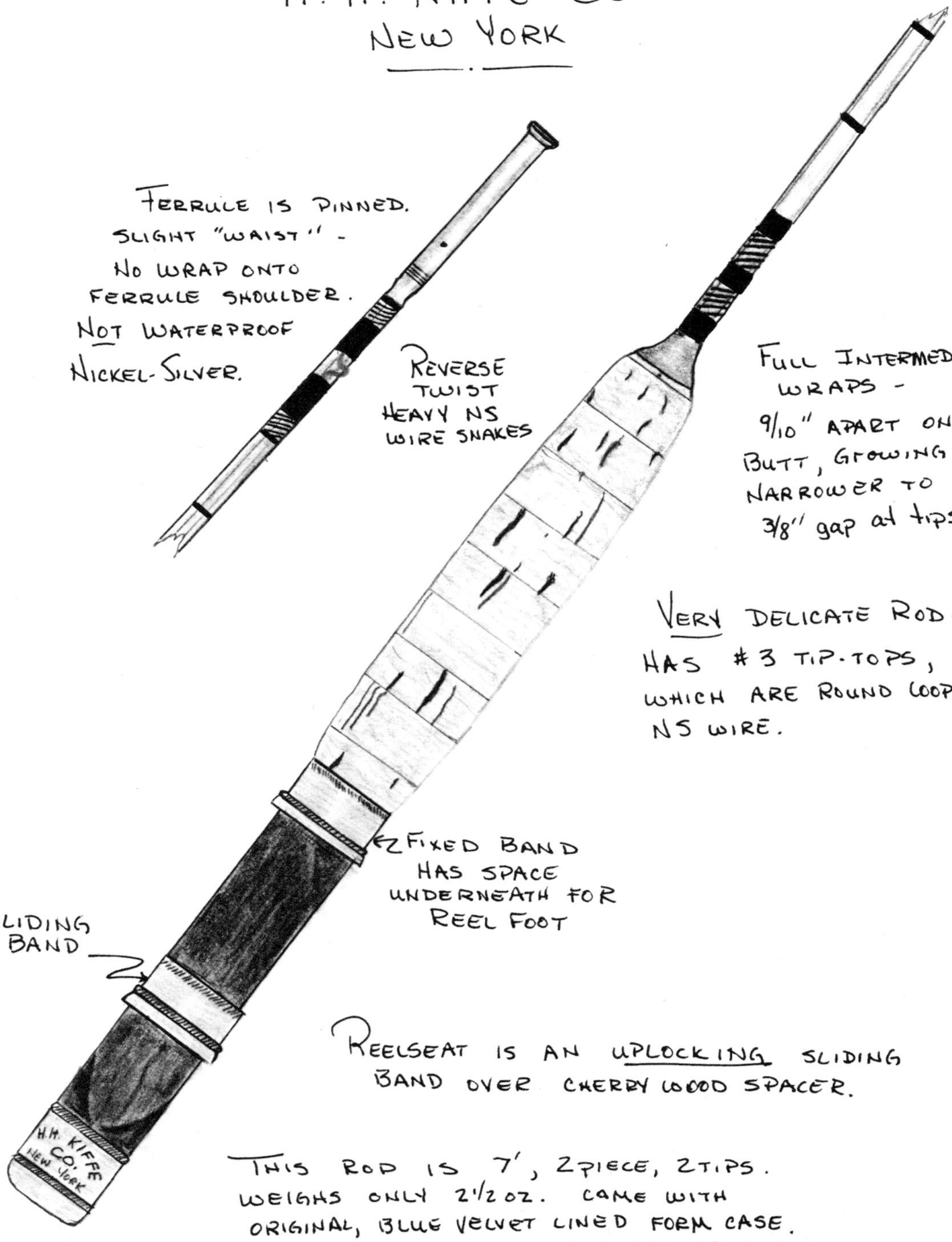

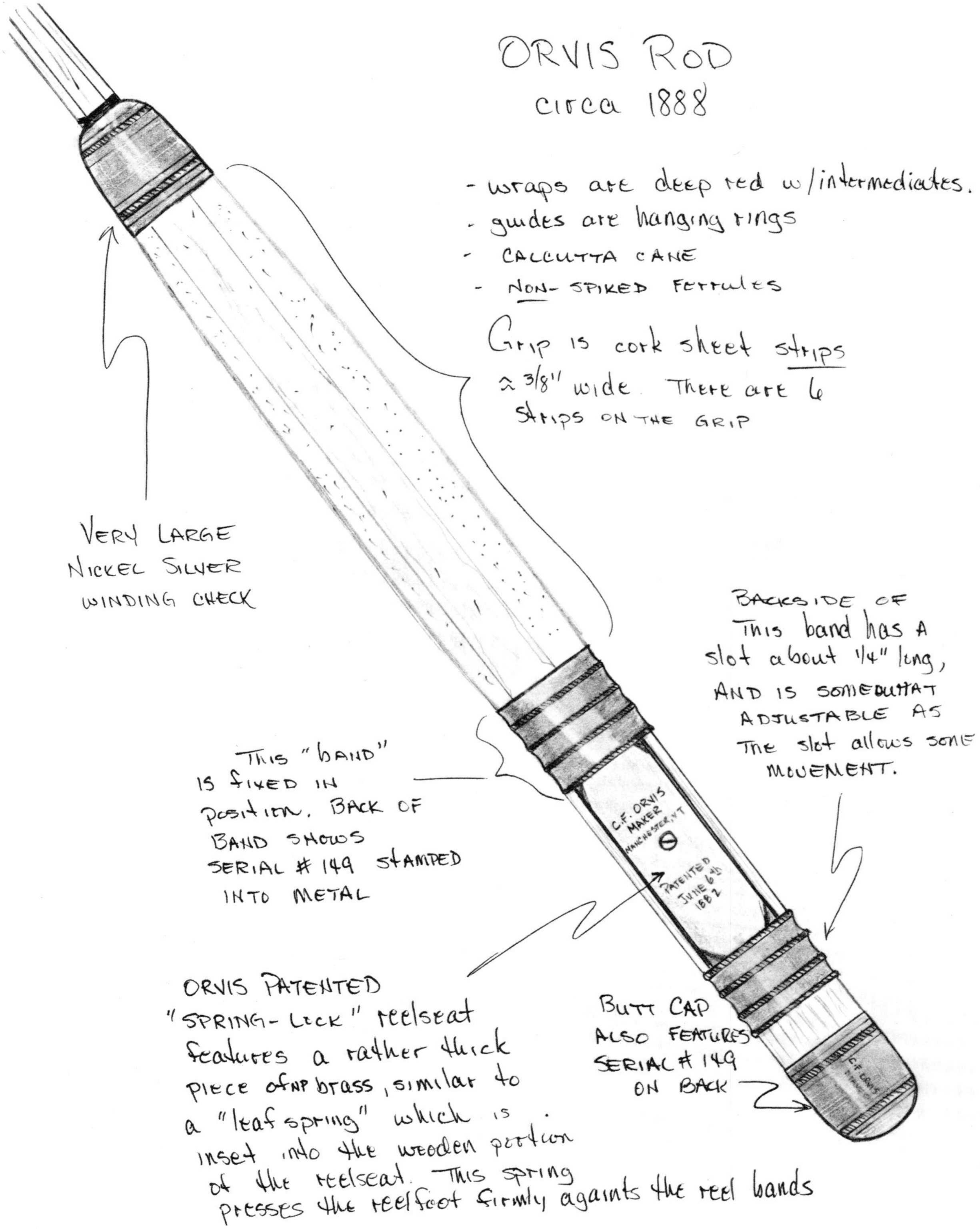
ORVIS ROD
circa 1888
- wraps are deep red w/intermediates.
- guides are hanging rings
- CALCUTTA CANE
- NON- SPIKED Ferrules
Grip is cork sheet strips ≈ 3/8" wide. There are 6 strips on the grip
VERY LARGE NICKEL SILVER WINDING CHECK
BACKSIDE OF This band has A slot about 1/4" long, AND IS SOMEWHAT ADJUSTABLE AS The slot allows some MOVEMENT.
This "band" is fixed in position. BACK OF BAND SHOWS SERIAL # 149 stamped INTO METAL
C.F. ORVIS MAKER MANCHESTER, VT
PATENTED JUNE 6th 1882
ORVIS PATENTED "SPRING-LOCK" reelseat features a rather thick piece of NP brass, similar to a "leaf spring" which is inset into the wooden portion of the reelseat. This spring presses the reelfoot firmly againts the reel bands
BUTT CAP ALSO FEATURES SERIAL # 149 ON BACK

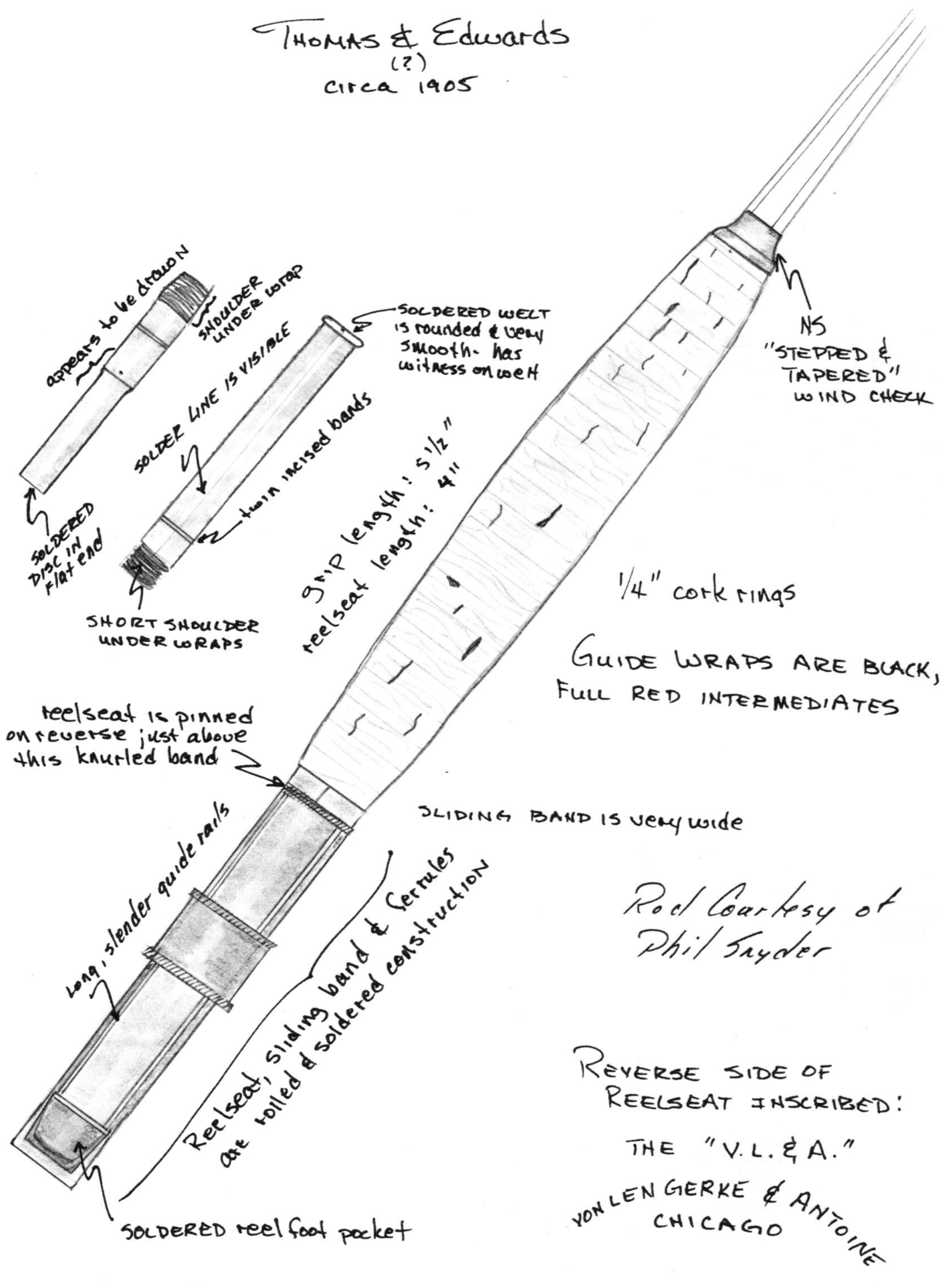

THOMAS & Edwards
(?)
circa 1905
appears to be drawn
SHOULDER UNDER wrap
SOLDERED WELT is rounded & very smooth. has witness on welt
SOLDER LINE IS VISIBLE
twin incised bands
SOLDERED DISC IN FLAT END
SHORT SHOULDER UNDER WRAPS
grip length: 5 1/2"
reelseat length: 4"
NS
"STEPPED & TAPERED" WIND CHECK
1/4" cork rings
GUIDE WRAPS ARE BLACK, FULL RED INTERMEDIATES
reelseat is pinned on reverse just above this knurled band
SLIDING BAND IS very wide
long, slender guide rails
Reelseat, sliding band & ferrules are rolled & soldered construction
Rod Courtesy of Phil Snyder
REVERSE SIDE OF REELSEAT INSCRIBED:
THE "V.L. & A."
VON LENGERKE & ANTOINE
CHICAGO
SOLDERED reel foot pocket

The Fred D. Divine Company

One of the oldest and longest lived rod companies was founded in about 1875 in Utica, New York. Fred Divine was a well-educated entrepreneur with many interests. He truly enjoyed fishing and hunting, and his rod business was a result of that love of the outdoors. Fred made all of the rods himself for the first few years, but by 1888 he had a small operation with several employees.

In 1885, Fred Divine patented a reelseat that was used for about 40 years. This reelseat is usually seen on Lancewood and Greenheart rods, and very seldom on cane rods. Divine also patented a mildly spiralled rodshaft in 1892, and these were made for about ten years. A new factory was opened in 1898, and the production of rods was increased threefold.

In 1900, the Divine rod factory was the victim of a fire. Though the fire itself was not a disaster, Fred Divine was killed in the process of inspecting the damage on the following day. His widow took over the business with the help of her brother, George McDuffee, who had been working with Fred since 1888. In 1904, The Divine Special was introduced. Fred Divine had designed this rod to suit his own tastes, and Mrs. Divine added it to the line in his memory.

From 1900 to 1919, the company made thousands of cheap trade rods, many of them of rounded Calcutta cane. These were often sold by Abbey & Imbrie as *Steel Vine* rods, and under many other trade names by most large sporting goods houses. These were never marked or sold as Divine Rods. High quality Divine Rods were still being made, though in smaller numbers, and were sold through outlets such as Abercrombie & Fitch.

After 1900, all Divine Rods carried a serial number. From 1900 to 1909, it was simply a series of numbers. The easiest way to spot a Divine Rod is the reelseat. If the reelseat was wood, it was turned at the grip in two steps to form a wood grip check. No other maker has ever used this style of reelseat. The wood spacer was mortised to receive the foot of the reel. Metal reelseats were of several styles and always had the serial number stamped into the seat, such as "R2460". The letters are useful for determining the relative age.

The letter was added in 1910, and "A" indicates that year, "B" is 1911, and so on. For example, a rod marked "Axxx" was earlier than one marked "Hxxx". After 1917, a decal was placed on the rodshaft above the hookkeeper.

The most common Divine Rod was the Standard Trout Model. See the drawing for details. This rod used no winding check, but had the forward end of the grip tapered down to the cane and wrapped over with silk. All Divine Rods came equipped with a hookkeeper, which was wound in the open style. (An *Open* hookkeeper wrap has a gap between the feet of the hookkeeper bridle, or strap.)

At the close of WW II, George McDuffee bought out his sister's interest in the company. His head rodmaker was Frank Becraft, who had been with Fred Divine since 1892. Frank Becraft was responsible for most of the design features and the tapers of Divine Rods. In 1917, he introduced the finest of the Divine Rods, The Fairy Fly Rod. See the drawing for details. Through the remaining years of production, Becraft was the rodmaker, and McDuffee looked after the business end of the operation.

Many new models were introduced in 1920, including the Rainbow, Special Dry Fly, and Travelling Man's Friend. Later additions were the Golden West (later called Gloriwest) in 1923, and in 1927, the Celdimoc, Pathfinder, Echo, and Peacemaker. The Rainbow model was essentially the same as the Standard Trout rod, but was more *customized*. It could be ordered with dark or light cane and with any wrap colors the customer wanted. There were a couple of standard combinations for the silk colors. These are listed in the wrap spec chart.

The Great Depression was essentially the end of the Fred D. Divine Company, though the company was not shut down. Frank Becraft retired in 1932, and McDuffee was unable to keep the plant operating very long without his help. In the years just before WW II, the rodmaking equipment and supplies were sold to the Horrocks-Ibbotson Company, which was also manufacturing rods in Utica.

For dating rods, the markings and decals are very useful. Rods marked "The / Divine / Rod" on the metal reelseat are pre World War I. From 1916 to 1926, the rods had only the serial number on the reelseat, and a rectangular decal was placed at an angle on the rodshaft above the hookkeeper wrap. This decal was very simple with bold block style lettering in black on a white background. From late 1926 until 1928, the decal was the same, but the lettering was script style. In 1928 the *fancy* decal came into use on the higher grades and the plain decal was used on the lowest grade, the Pathfinder. The *fancy* decal is gold foil, with the border and script lettering in black. The Pathfinder decal is simple black block letters on white.

Guides used on Divine rods were usually seen in consistent pairings. For example; rods with Agate (A) strippers had English (E) reverse-twist snake guides; those with Perfection (Perf) single ring guides used American (Am) twist perfection snakes. In the Hardware Specifications charts this is shown: *A/Perf* for strippers and *E or Am* for snakes. The Fairy Fly Rod was equipped with a tiny agate guide until the mid-1920's when the stripper was changed to a Perfection snake guide.

THE FRED D. DIVINE CO.

505 ROBERTS ST. UTICA, N. Y.

THE
DIVINE ROD
UTICA, N. Y.

The
Divine Rod
Utica, N.Y.

The
Divine Rod
Utica, N.Y.

DIVINE
PATHFINDER

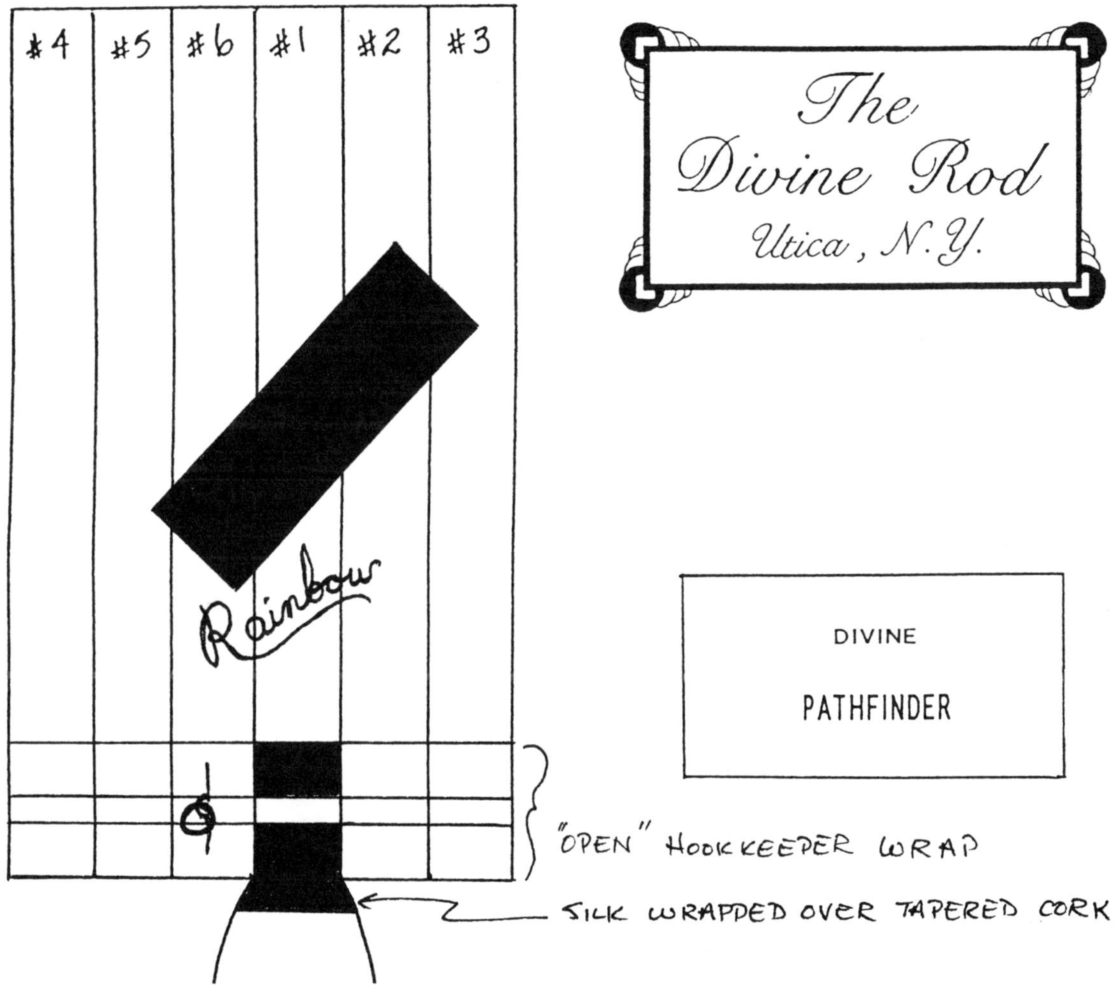

The Fred D. Divine Co. **Hardware Specs** **1920/1930's**

Model	Reelseat	Grip	Stripper	Snake	TT	Ferrules
Standard Trout	Choice	Cigar	Agate	E, Steel	P	NS, HW, S
Rainbow	Choice	Cigar	A/Perf	E, Steel	P	NS, HW, S
Divine Special	C&R Cedar	Mild Full Wells	Agate/Perf	E, Steel	P	NS, HW, S
Special Dry Fly	English	Straight	Agate	E, Steel	A	NS, HW, S, WP
Fairy Fly Rod	C&R Cork	Cigar	A/Snake	E, Steel	P	NS, HW, S. WP
Invisible Rod	Choice	Cigar	A/Perf	E, Steel	P	BNS, HW, S
Travel' Man Friend	NS, SB	Straight	Perf	E, Steel	P	NS, HW, S
Gloriwest	SDL, Black	Rev Half Wells	Perf	Am, Steel	P	BNS, HW, S
Celdimoc	Choice	Half Wells	Perf	Am, Steel	P	BNS, HW, S
Peacemaker	Choice	Straight	Perf	Am, Steel	P	BNS, HW, S
Pathfinder	Choice	Cigar	Perf	Am, Steel	P	NS, HW, S
Echo	Choice	Cigar	Perf	Am, Steel	P	NS, HW, S
Fly-Wate	C&R Cork	Cigar	Perf	Am, Steel	P	NS, HW, S, WP

The Fred D. Divine Co. — Wrap Specs — 1920/1930's

Model	Main Wraps	Tipping Wraps	Intermediates	
Standard Trout	Red #11	Black #1	Red #11	
Rainbow - light cane	Red #10 or B/W #X	Yellow #5 or Red #10	Red #10 or B/W #X	
Rainbow - dark cane	Lime #19 or Red #10	Red #10 or Lime #19	Lime #19 or Red #10	
Divine Special	Red #11	Black #1	Red #11	
Special Dry Fly	Red #11	None	Red #11	
Fairy Fly Rod	Red #11	None	Red #11	
Invisible Rod	Olive #21	None	Olive #21	
Travel' Man Friend	Red #11	Black #1	Red #11	
Gloriwest	Green #25	None	Yellow #5	
Celdimoc	B&W Japser #X	Yellow #5, both ends	Yellow #5	
Peacemaker	Red #10	Olive #21	Red #10	above last
Pathfinder	Green #25	Red #10	Red #10	snake guide
Echo	Black #1	None	Black #1	on tips only
Fly-Wate	Clear (Yellow #1)	Black (if any)	None	

Divine Standard Trout

c. 1922

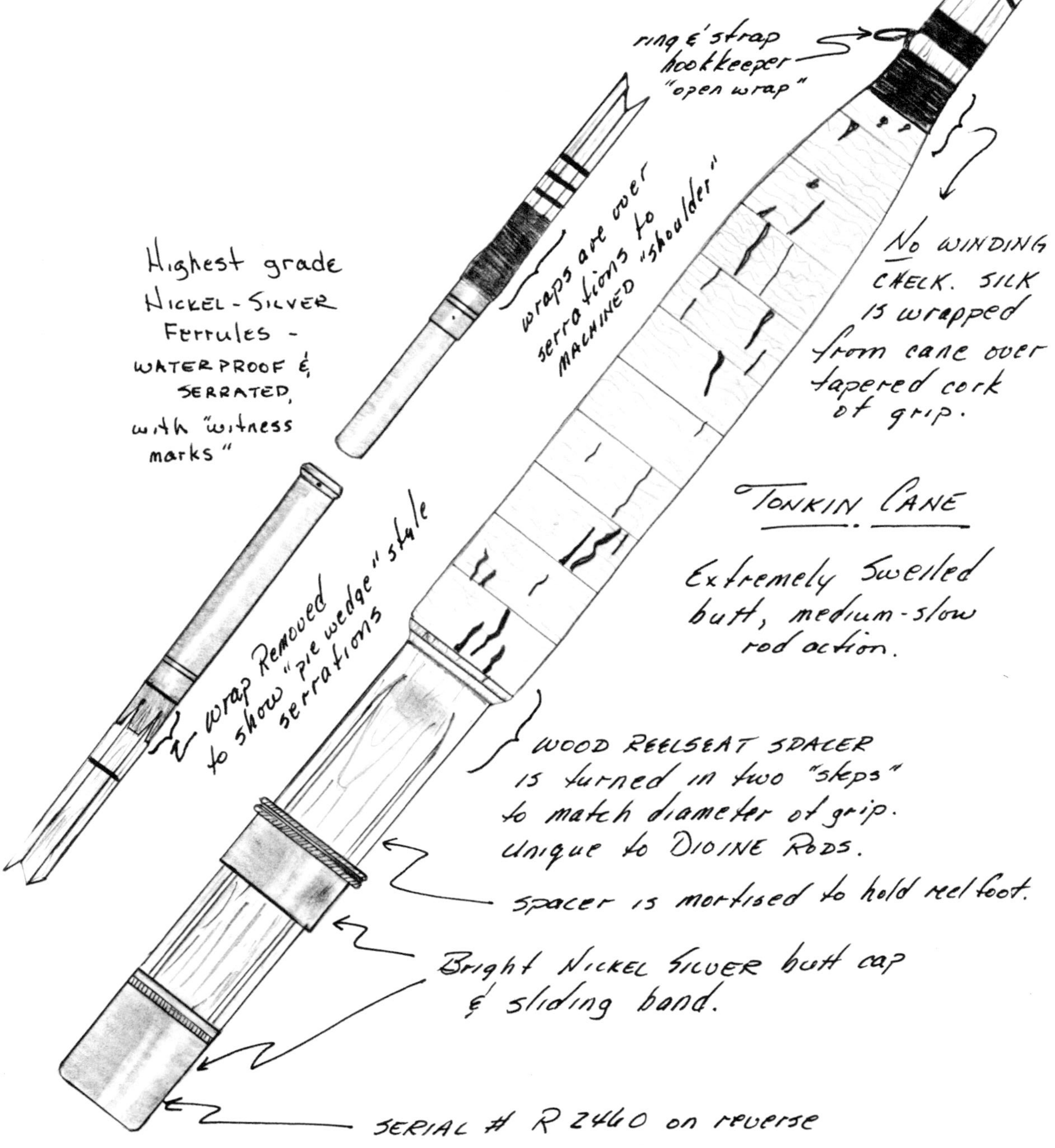

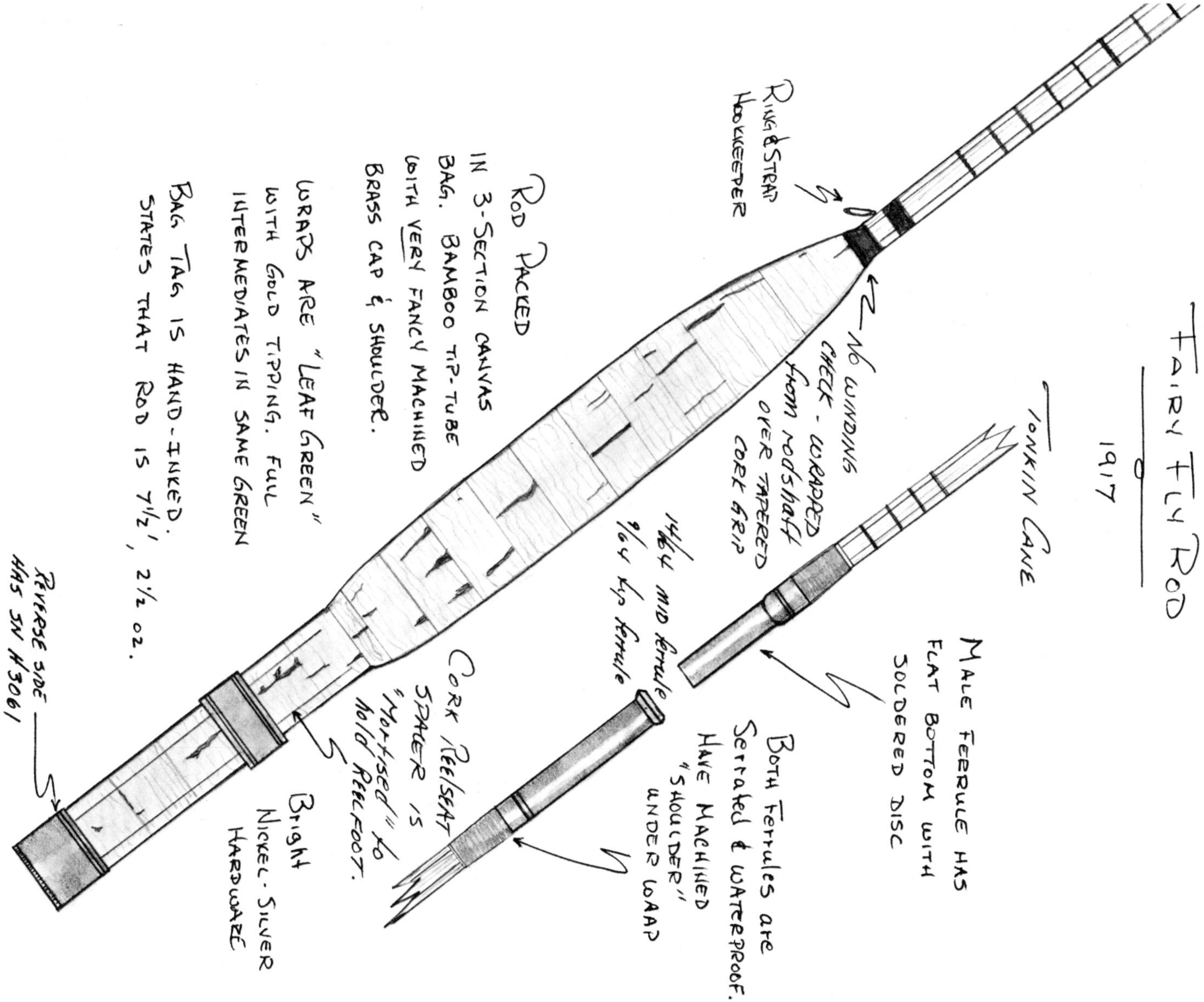
FAIRY FLY ROD
1917
TONKIN CANE
Ring & Strap Hookkeeper
No winding check - wrapped from rodshaft over tapered cork grip
14/64 mid ferrule
9/64 tip ferrule
MALE FERRULE HAS FLAT BOTTOM WITH SOLDERED DISC
Both ferrules are serrated & waterproof. Have machined "shoulder" under wrap
ROD PACKED IN 3-SECTION CANVAS BAG. BAMBOO TIP-TUBE WITH VERY FANCY MACHINED BRASS CAP & SHOULDER.
WRAPS ARE "LEAF GREEN" WITH GOLD TIPPING. FULL INTERMEDIATES IN SAME GREEN
BAG TAG IS HAND-INKED. STATES THAT ROD IS 7½', 2½ oz.
Cork reelseat spacer is "mortised" to hold reelfoot.
Bright Nickel-Silver Hardware
Reverse side has SN #3061

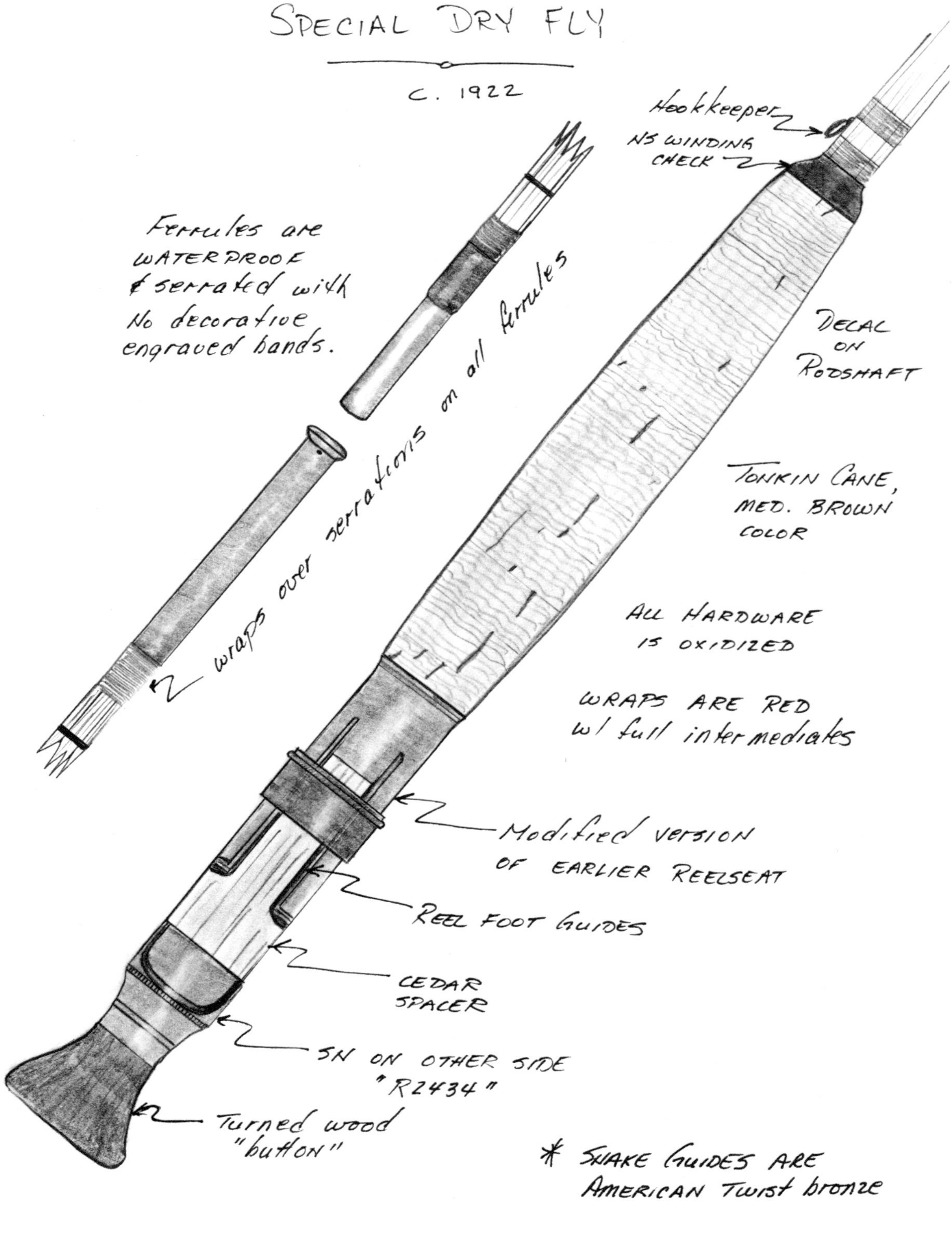
SPECIAL DRY FLY
C. 1922
Hookkeeper
NS WINDING CHECK
Ferrules are WATERPROOF & serrated with No decorative engraved bands.
wraps over serrations on all ferrules
DECAL ON RODSHAFT
TONKIN CANE, MED. BROWN COLOR
ALL HARDWARE IS OXIDIZED
WRAPS ARE RED w/ full intermediates
Modified version OF EARLIER REELSEAT
REEL FOOT GUIDES
CEDAR SPACER
SN ON OTHER SIDE "R2434"
Turned wood "button"
* SNAKE GUIDES ARE AMERICAN Twist bronze

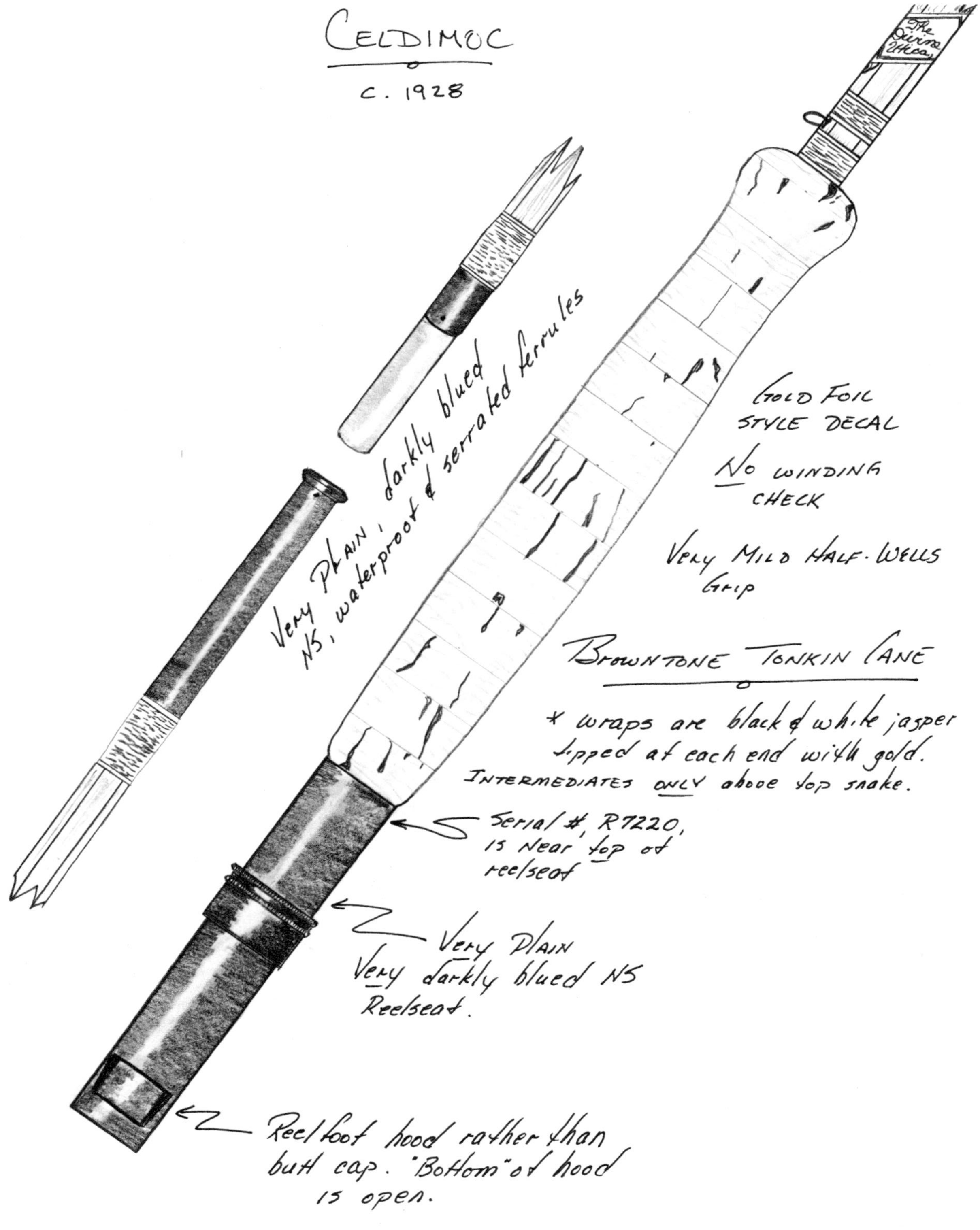
CELDIMOC
c. 1928
Very Plain, darkly blued NS, waterproof & serrated ferrules
Gold Foil Style Decal
No winding check
Very Mild Half-Wells Grip
Browntone Tonkin Cane
* wraps are black & white jasper tipped at each end with gold.
Intermediates only above top snake.
Serial #, R7220, is near top of reelseat
Very Plain
Very darkly blued NS Reelseat.
Reelfoot hood rather than butt cap. "Bottom" of hood is open.

Goodwin Granger Company

Granger Rods were catalogued using a different terminology than that of most makers. In Granger's lexicon, the *Grade* of the rod was its quality level designated by a name such as Special or Favorite. The *Model* represented the length and weight class of each rod within a Grade, and was written as the initials of the Grade name with the length and weight designators; for example : GF9050 = Granger Favorite, 9'0", 5.0 oz.

These were never written on the rod, but appeared only on the paper label attached to the rod's aluminum tube. Without the label, the only way to tell the model is to refer to the chart on page 81. Measure the two ferrule sizes and compare to the chart.

Granger rods fall neatly into three historic periods, and are very easy to date within five years or less. The earliest rods are rarely seen, and not many were made. See the drawing on page 85. These rods were made from c. 1918 to c 1924.

The *Middle-Era* (c. 1924 to c. 1929) rods are more easily found. The earliest of these is marked with style #1 on the reelseat, and with the Grade name engraved between the knurled bands of the reelseat. After early 1928, marking #2 was used.

The *Late-Era* Granger rods are those most commonly seen. These have the famous Granger Patent, all NS, uplocking, internal thread reelseat and a mild reversed half-Wells grip. Refer to the flat map and markings chart.

The reelseat is roll stamped with marking #2, and from 1934 to 1938, marking #3 appeared above the maker's mark. After April of 1938, Marking #4 was roll-stamped onto the reelseat above the maker's mark. The Grade name designation was rubber stamped on the #1 flat just above the wind check wrap, with the printing running from the grip toward the ferrule. Some of these are faint and quite difficult to read. Exercise great caution in trying to save this grade designation! It is the only provenance you will have for a high grade Granger. This is because all Granger rods used the same grip style, reelseat, winding check and ferrules, regardless of grade.

The only real concession to cost was in the grading of the cork rings used in the grip, the cane work (node mismatching, etc.) and the number of guides and degree of artistry expended in wrapping. Even the lowest grade (Champion) had trim wraps and enough guides to perform well.

Granger rods use a 3x3 node spacing. Rods that had the best spacing and clearest cane were designated the higher grades, and those with less than perfect spacing were down-graded accordingly. Granger's cane was ammonia treated to give it a rich caramel color. Only two rod models, the Denver Special and the Colorado Special, were made of light cane. These light cane Grangers are fairly scarce, especially in the 8½' models.

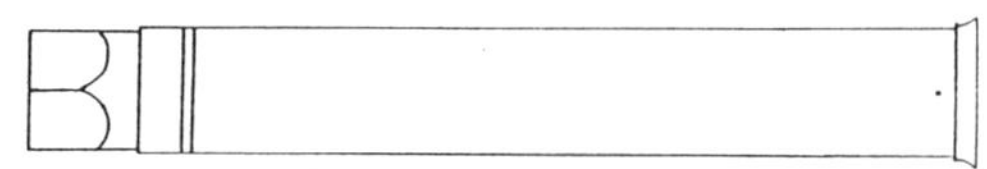

Granger Ferrule – Actual Size

Ranking of the Granger grades, from the highest to the lowest:

Early Era	*Middle Era*	*Late Era*
Tournament	The Granger Rod	Registered
Premier	The Goodwin Rod	Premier
DeLuxe	Denver Special	DeLuxe
Special	Colorado Special	Favorite
		Aristocrat
		Special
		Victory
		Champion

Manufactured by

GOODWIN GRANGER COMPANY

Forest Drive at Grant — Denver, Colorado

Granger Rods

#1

MADE BY
GOODWIN GRANGER & CO.
DENVER

#2

MADE BY
GOODWIN GRANGER CO.
DENVER

#3

PAT. PEND.

#4

PAT. APR. 12, 1938
PAT. NO. 2,114,107

GRANGER ROD MODELS

MODEL	LENGTH	WEIGHT	BUTT FERRULE	MID FERRULE	LINE SIZE
7030	7'0", 2/2	3 OZ.	12	N/A	HEH
7633	7'6", 3/2	3 3/4	15	9	HEH
8040	8'0", 3/2	4 OZ.	16	10	HEH
8642	8'6", 3/2	4 1/2	17	11	HEH
9043	9'0", 3/2	4 3/4	17	11	HDH
9050	9'0", 3/2	5 OZ.	17	13	HDH
9053	9'0", 3/2	5 3/4	17	14	HCH
9660	9'6", 3/2	6 OZ.	18	15	HCH
1062	10'0", 3/2	6 1/2	19	16	GBG

Goodwin Granger Company — Hardware Specs — 1920's - 1939

Era	Grade Name	Reelseat	Grip	Stripper	Snakes	TipTops
1920's	Colorado Special	All N/S		S. Snake	N/S	M, HS
	Denver Special	Sliding Band		S. Snake	N/S	M, HS
	The Goodwin Rod	with		Agate	TS	M, TS
	The Granger Rod	Soldered Hood	Coke Bottle	Agate	TS	M, TS
1930	Granger Special	After 1930		PC	HS	PC
	Granger Favorite	Grade Name		PC	HS	PC
	Granger DeLuxe	Engraved		PC	HS	PC
	Granger Premier	Between Knurls		PC	PC	PC
1934	Granger Champion	N/S Slide Band		Garnix	HS	M, CPNS
	Granger Victory			PC	HS	PC
	Granger Special			PC	HS	PC
	Granger Aristocrat	Granger Patent	Mild	PC	HS	PC
	Granger Favorite	N/S Uplocking	Reversed	PC	HS	PC
	Granger DeLuxe	Internal Thread	Half–Wells	PC	HS	PC
	Granger Premier			PC	PC	PC
	Granger Registered			PC	PC	PC
1939	Granger Two Piece	N/S Slide Band	Hammer	PC	HS, PC	PC

Granger Wrap Specs #1

Colorado Special
Red #10
Gold #7 - in signature only

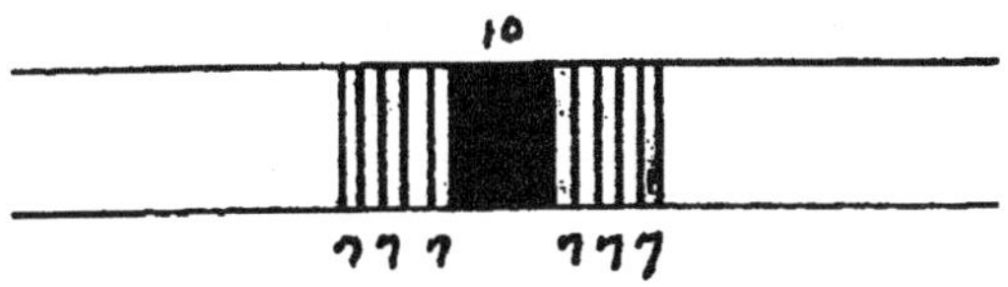

Mid Butt Trim

Denver Special
Red #10
Gold #7 - in signature only
Red #10 for Intermediates

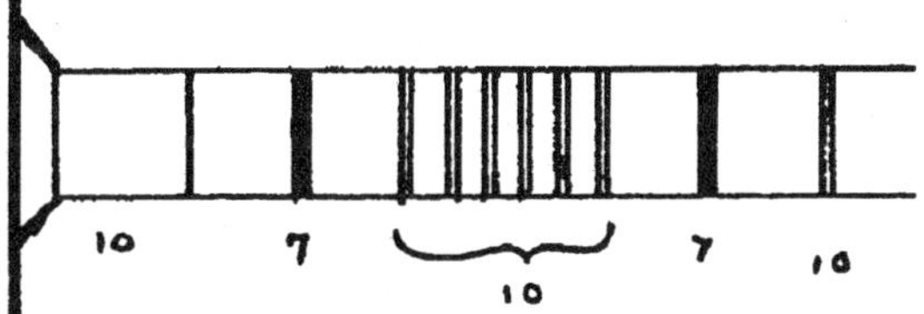

Signature Wrap

The Goodwin Rod
B&W Jasper #
Gold #7 for Tipping
and Intermediates

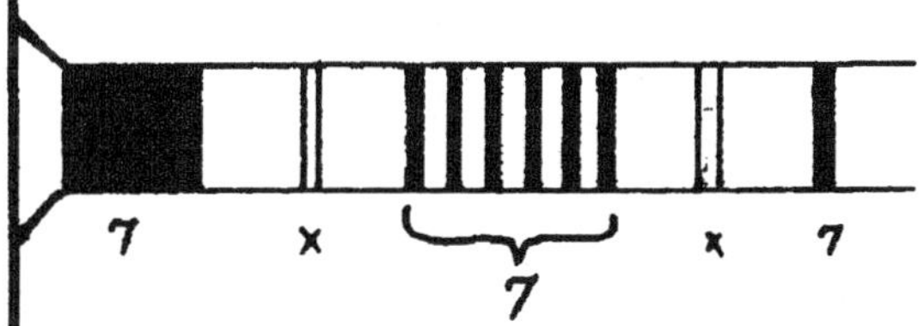

Signature Wrap

The Granger Rod
B&W Jasper #
Gold #7 for Tipping
and triple border wraps

Mid Butt Trim

Granger Special
Granger Green #20
Plain

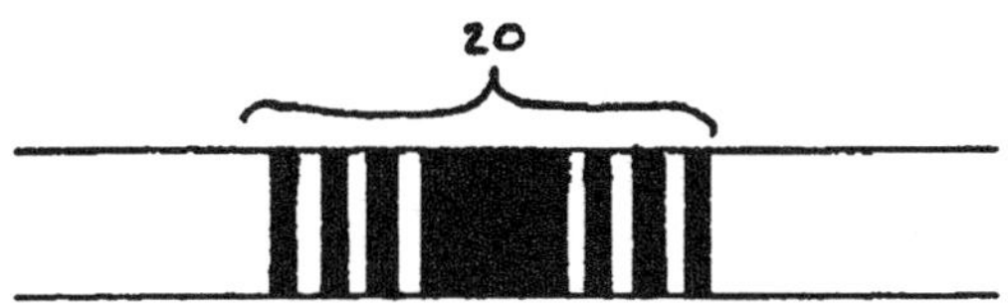

Mid Butt Trim

Champion
Red #10
Black #1 for Tipping

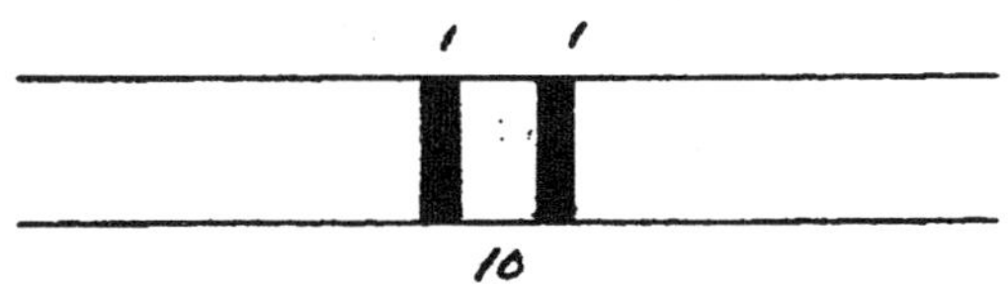

Mid Butt Trim

Granger Wrap Specs #2

Victory
B&O Jasper #XX
Tipped Black #1

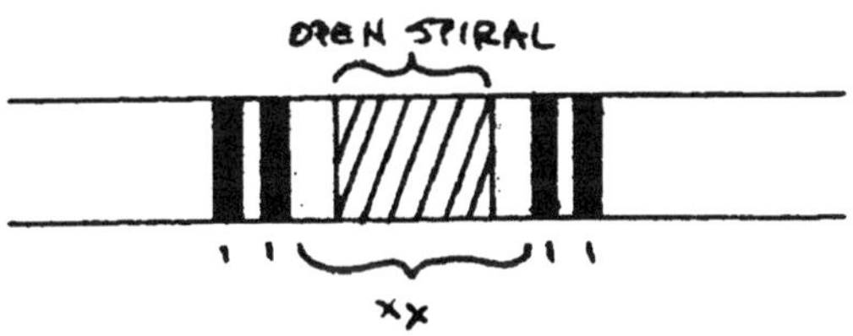

Mid Butt Trim

Aristocrat
Beige #27
Tipped Brown #28

No Signature or Mid Butt Wrap

Favorite
B&W Jasper #X
Gold #7 for Tipping

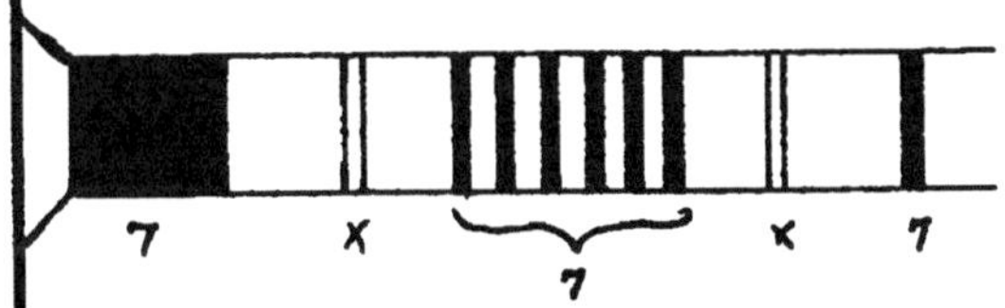

Signature Wrap

DeLuxe
B&W Jasper #X
Gold #7 for Tipping
with 3 border wraps Gold #7

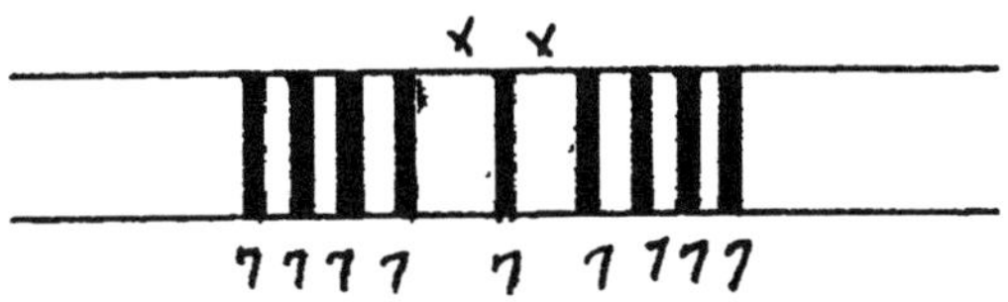

Mid Butt Trim

Premier
Gold #7
Plain - No Tipping

No Signature or Mid Butt Trim

Registered
Black #1
Tipped White #2
at each end of each wrap

No Signature or Mid Butt Trim

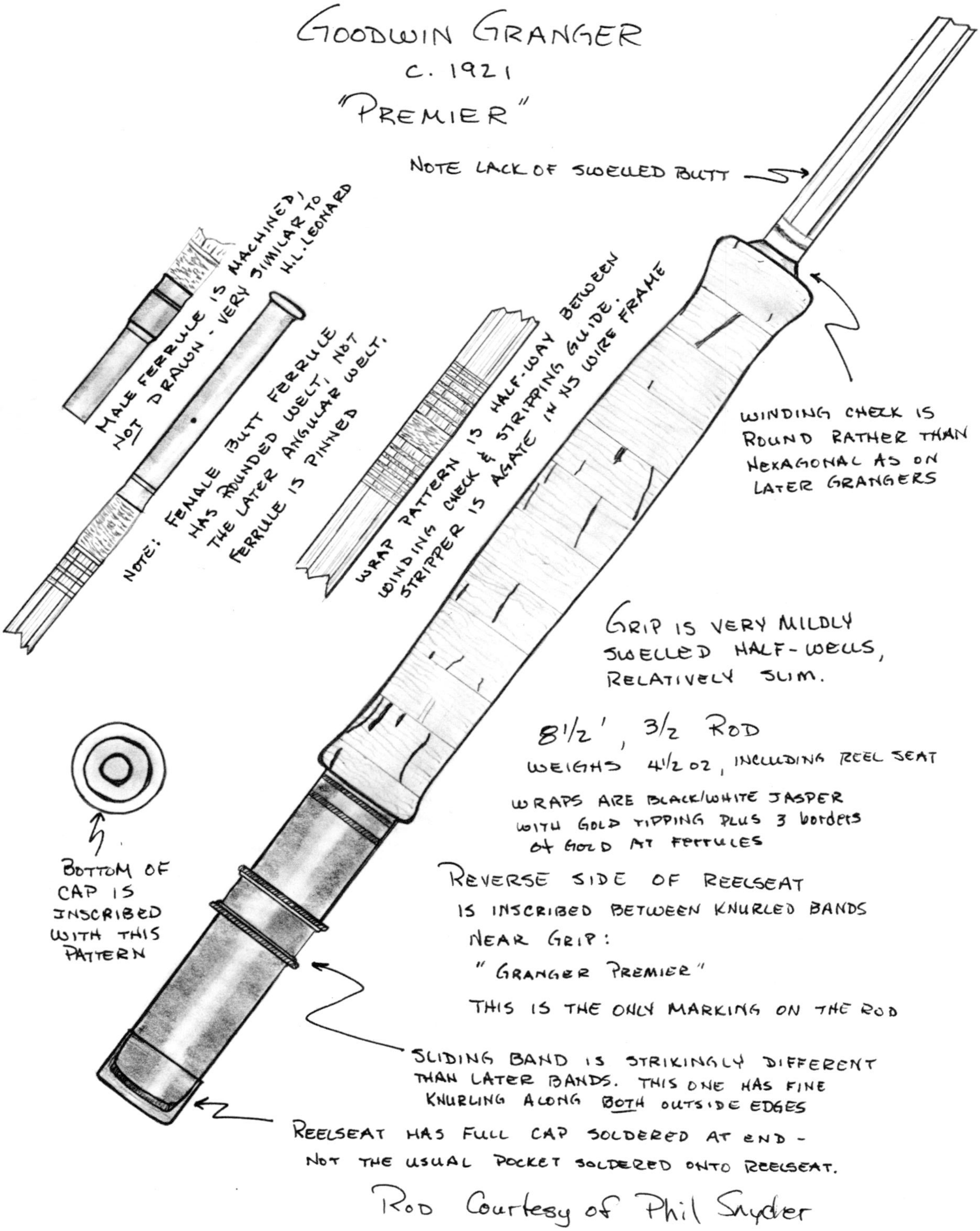
GOODWIN GRANGER
C. 1921
"PREMIER"
NOTE LACK OF SWELLED BUTT
MALE FERRULE IS MACHINED, NOT DRAWN - VERY SIMILAR TO H.L. LEONARD
NOTE: FEMALE BUTT FERRULE HAS ROUNDED WELT, NOT THE LATER ANGULAR WELT. FERRULE IS PINNED
WRAP PATTERN IS HALF-WAY BETWEEN WINDING CHECK & STRIPPING GUIDE. STRIPPER IS AGATE IN NS WIRE FRAME
WINDING CHECK IS ROUND RATHER THAN HEXAGONAL AS ON LATER GRANGERS
GRIP IS VERY MILDLY SWELLED HALF-WELLS, RELATIVELY SLIM.
8 1/2', 3/2 ROD
WEIGHS 4 1/2 OZ, INCLUDING REEL SEAT
WRAPS ARE BLACK/WHITE JASPER WITH GOLD TIPPING PLUS 3 BORDERS OF GOLD AT FERRULES
BOTTOM OF CAP IS INSCRIBED WITH THIS PATTERN
REVERSE SIDE OF REELSEAT IS INSCRIBED BETWEEN KNURLED BANDS NEAR GRIP:
"GRANGER PREMIER"
THIS IS THE ONLY MARKING ON THE ROD
SLIDING BAND IS STRIKINGLY DIFFERENT THAN LATER BANDS. THIS ONE HAS FINE KNURLING ALONG BOTH OUTSIDE EDGES
REELSEAT HAS FULL CAP SOLDERED AT END - NOT THE USUAL POCKET SOLDERED ONTO REELSEAT.
Rod Courtesy of Phil Snyder

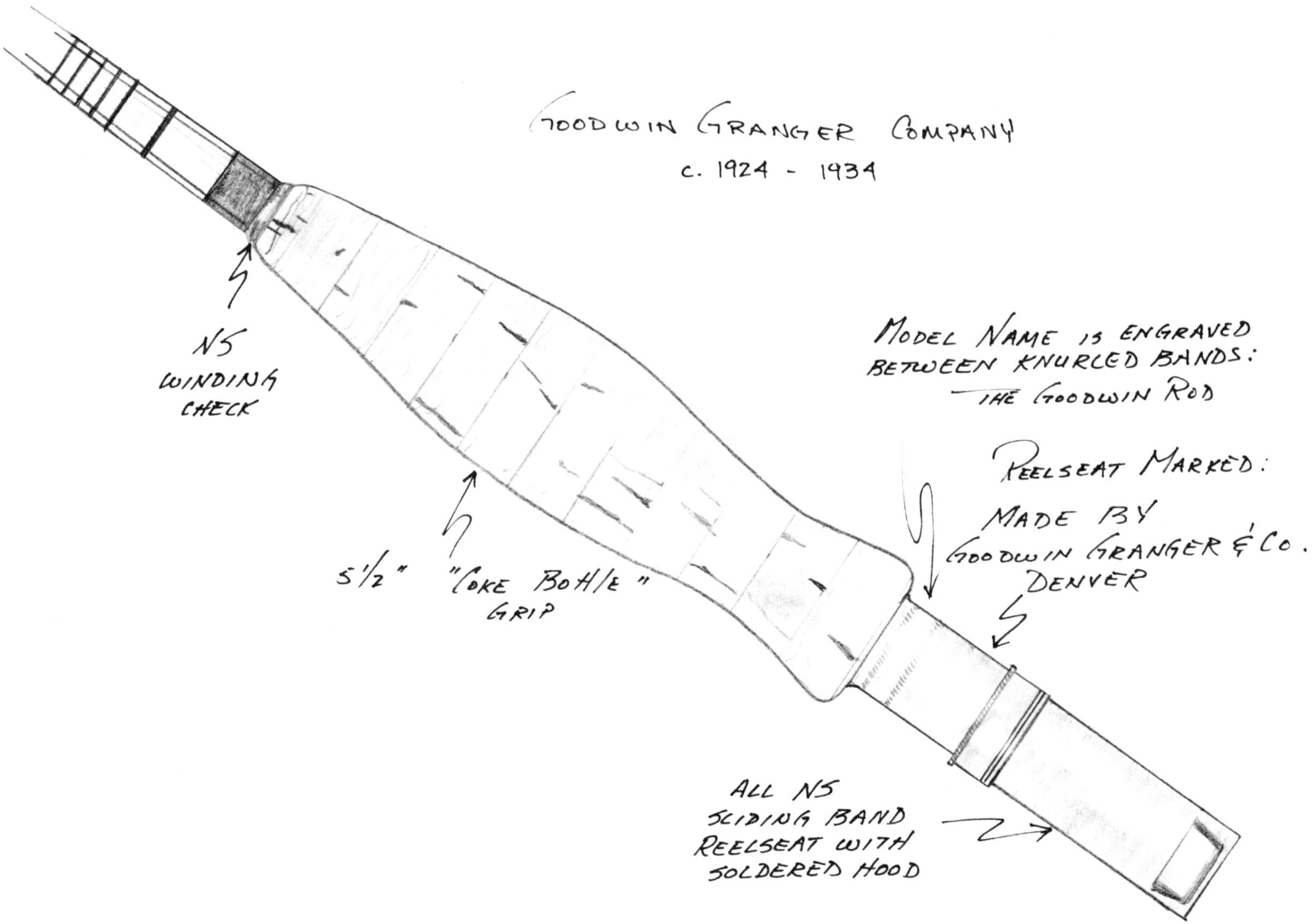
GOODWIN GRANGER COMPANY
c. 1924 - 1934
NS WINDING CHECK
5½" "COKE BOTTLE" GRIP
MODEL NAME IS ENGRAVED BETWEEN KNURLED BANDS: THE GOODWIN ROD
REELSEAT MARKED:
MADE BY GOODWIN GRANGER & CO. DENVER
ALL NS SLIDING BAND REELSEAT WITH SOLDERED HOOD

Goodwin Granger Company

1934 → 1941

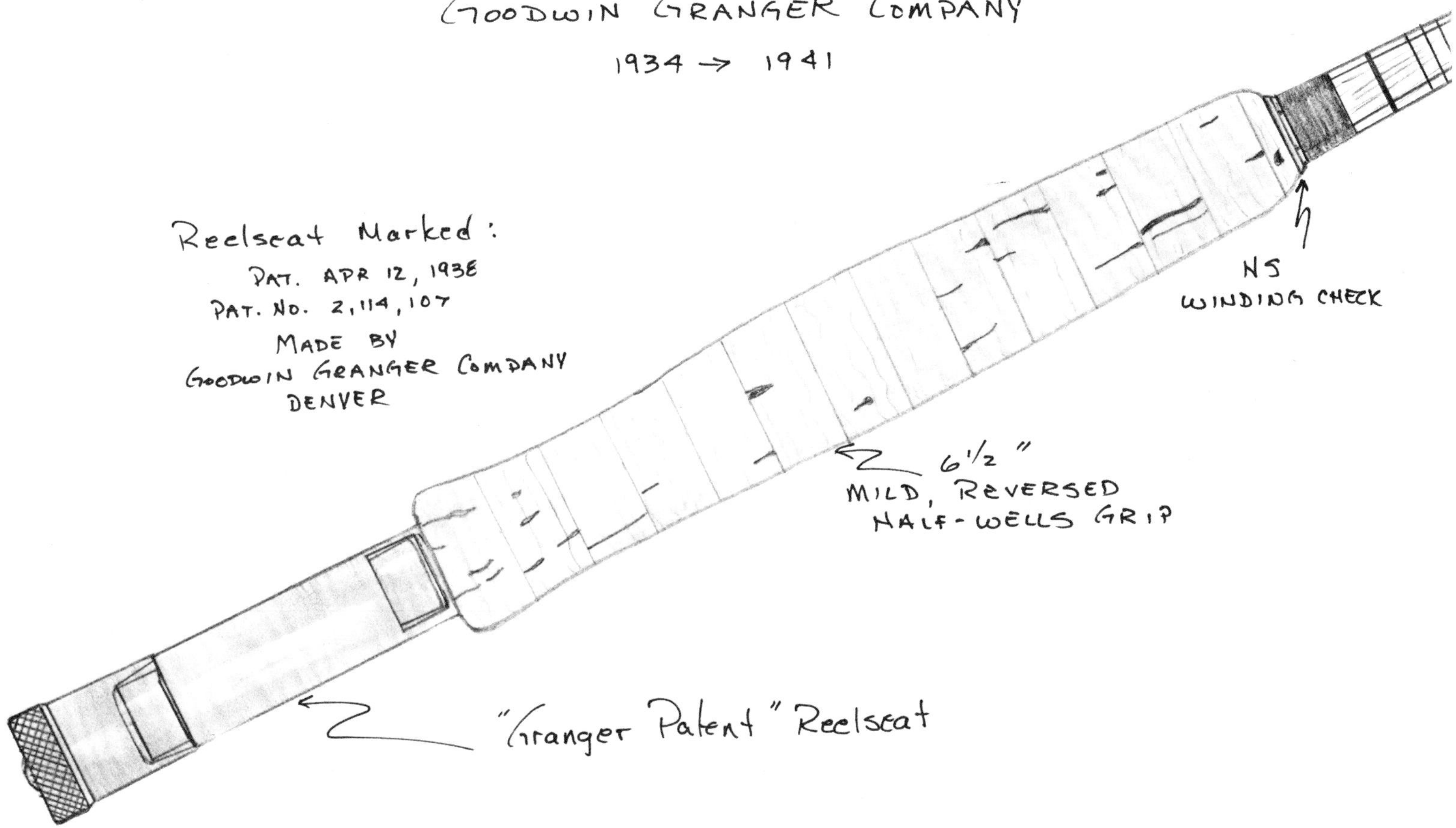

James Heddon's Sons

Heddon was the most prolific maker of better bamboo fly rods. It is estimated that from the 1930's through the 1940's, Heddon produced as many as 100,000 rods per year. The remarkable feature of all Heddon rods is the tremendously high quality standards that were *consistently* maintained. No other maker of so many rods was able to meet this achievement.

Although Granger and Phillipson turned out comparable quality rods, neither produced as many rods as Heddon, and though H. L. Leonard rods are almost as numerous as Heddons, the quality was inconsistent. Another amazing fact is that despite the huge number and consistent high quality of rods manufactured, the rodmaking business was incidental to Heddon, and occurred almost as an afterthought.

Heddon Made - Well Made was not just a slogan to the people of Dowagiac (pronounce this DOE-WAH'-JACK), Michigan. It is said that at one time or another, almost everyone in Dowagiac had a family member who was employed by Heddon. Heddon was a family business in the truest sense of the concept, but the family was extended to include every employee.

Heddon's first involvement with rodmaking was in the manufacture of bait rods. This was a natural extension of the bait-making business. The first bait rods under the Heddon label were made in about 1908. These rods were of blonde colored cane, had swelled butts and fancy-shaped grips made of narrow cork rings. Heddon pioneered the concept of the *Short Butt - Long Tip* design.

The Heddon Fly Rod was introduced in 1924. It was made in one grade only - the finest - and sold for $35. It was appropriately called the No. 35 for that reason. With the introduction of the fly rods, Heddon began using its famous *brown-toned Tempered Cane.* Two years later, the No. 20 was added to the line at a cost of $20. It was called *Bill Stanley's Favorite* from the very beginning in the catalogs, though it was much later that the name appeared on the rodhaft.

The No. 14 was added in 1927, and was different from the other rods is several ways. First is that its price did not match its model number as it sold for $15. Next, it had a grip made of narrow cork rings and a cap & ring reelseat with a gumwood spacer. Last, this rod was marked only #14 on the rodshaft, and had the same leaping Black Bass decal that was used on the bait rods.

In the late 1920's, Heddon added several more rod models and each model was finally given a name to go with its number.

Heddon rods of the 1920's are easily recognized by short, stubby grips and all NS sliding band reelseats. Usually the name Heddon is stamped on the bottom of the butt cap, and there are no other markings on the rods. The rod tube will have Heddon's Dowagiac stamped on the cap. The 1930's marked the beginning of Heddon's *classic era* of bamboo rod production. Several new rod Numbers were introduced. A new reelseat came into use. An earlier reelseat called Hold-Tite had been used on bait rods for several years and was modified to use on fly rods. In about 1932, the first true screw-locking seat was used on Heddon rods. It was still called the Hold-Tite but the seat was uplocking with Pyralin threads and a cedar spacer. During this period, the markings on the rod were as shown on the flat map on the far left of the page.

The next major change for Heddon rods came in 1939 with a new sytem of marking the rodshaft and a new reelseat. The markings were changed to those shown on the center flat map. These markings are the most commonly seen. The addition of the ferrule size and recommended line sizes is especially helpful to collectors who are trying to determine the action of a rod. The butt ferrule size is designated by "2F" or "2½F". The larger the number, the stiffer the rod. For three-piece rods, a 2F is usually a trout rod and 2½F is a bass rod. All two-piece rods are trout rods.

The new reelseat that was introduced at this time had both the thread and the spacer (barrel) of the reelseat made of Tenite. This was a down-locking style seat which usually featured blackened NS hardware. Some of these seats were made of black Pyaralin and were used on No. 17, Black Beauty rods. The tenite seats are usually a marbellized finish and most are brown, though the No. 14 used a Coral colored Tenite seat. The final changes for Heddon rods were only in the markings. In about 1951, the markings were changed to those seen on the far right side of the flat map page. Several rod models were dropped this year, but one new one was introduced, the No. 8 which was called simply The Heddon Fly Rod. It seems interesting to note that was the same name used when the first Heddon Fly rod was introduced 27 years earlier. Heddon not only made rods for their own catalog, but supplied many, many companies with rods to sell under their private label. Among the more common sellers of Heddon-made rods were Shakespeare, Weber, E.K. Tryon (Jay Harvey Rods and Kingfisher Rods) and Lyon & Coulson.

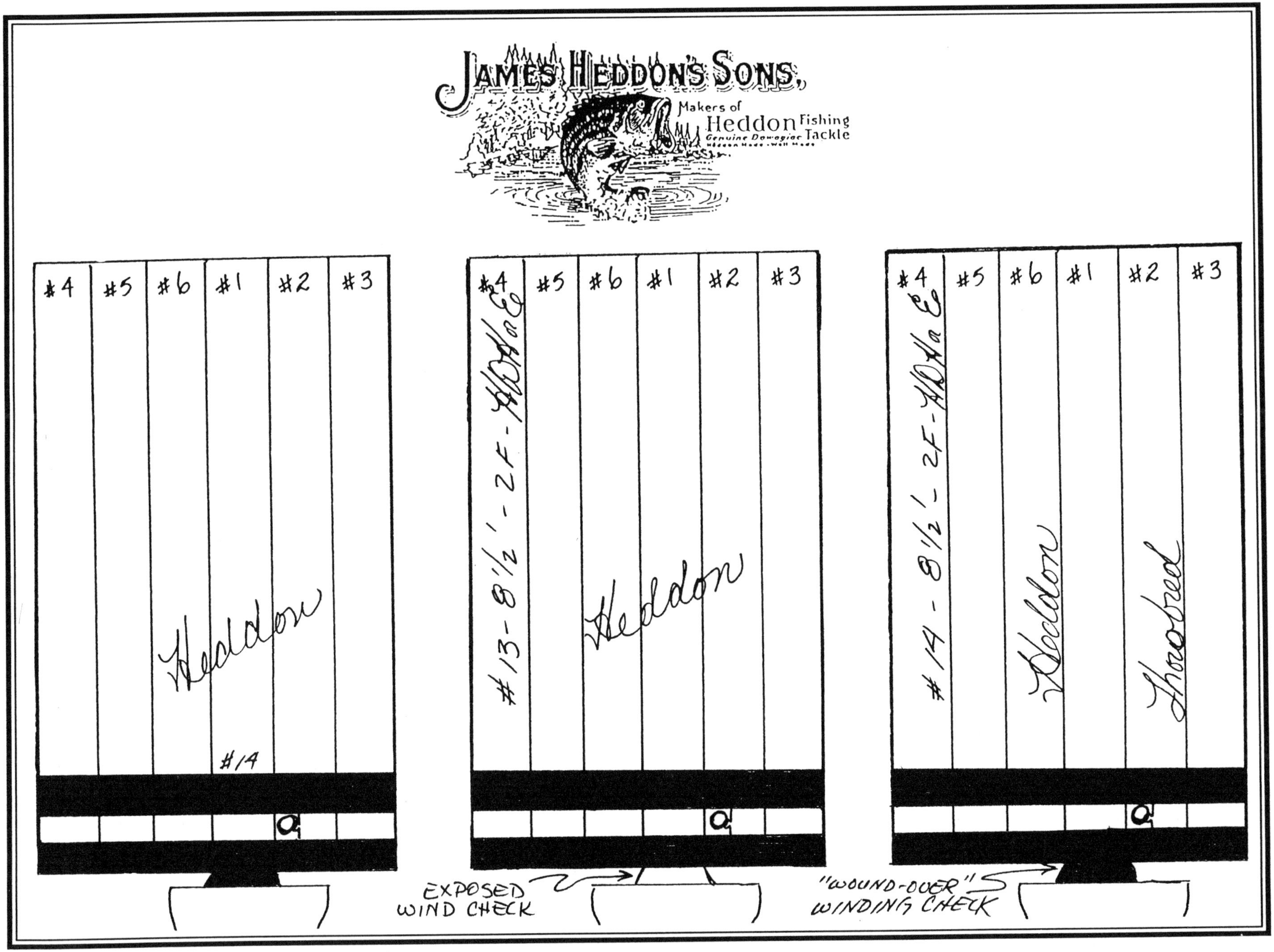

JAMES HEDDON'S SONS,
Makers of Heddon Fishing Tackle
Genuine Dowagiac
#4 #5 #6 #1 #2 #3
Heddon
#14
#4 #5 #6 #1 #2 #3
#13 - 8½' - 2F - HDHaE
Heddon
#4 #5 #6 #1 #2 #3
#14 - 8½' - 2F - HDHaE
Heddon
Thorobred
EXPOSED WIND CHECK
"WOUND-OVER" WINDING CHECK

James Heddon's Sons — Hardware Specs — 1924 - 1929

Year Intro	Rod #	Model	Reelseat	Grip	Stripper	Snake	TT	Ferrules
1927	14	No Name	C&R Gumwood	Straight - ¼" cork	Snake	TS	TS	Bright NS, RW
1926	20	Standard	C&R, NS	Half Wells	A	TS	TS	Bright NS, RW
1924	35		C&R, NS	Cigar	A	TS	TS	Bright NS, RW
1928	50	President	NS Screwtite	Full Wells	A	TS	TS	Bright NS, RW
1928	51	Princess 2/2	BNS, C&R, Cork	Cigar	A	PSS	PC	BNS, RW

James Heddon's Sons — Wrap Specs — 1924 - 1929

Rod #	Model	Wrap Color	Tipping Color	Intermediates & Trim
14	No Name	Red #10	N/A	Red #10, widely spaced
20	Standard	B&W Jasper #X	Yellow #3	
35		Yellow #3	Black #1	Yellow #3 - fancy Y/B signature
50	President	Silver/Grey #31	Black #1	
51	Princess	Silver/Grey #31	Black #1	

JAMES HEDDON'S SONS

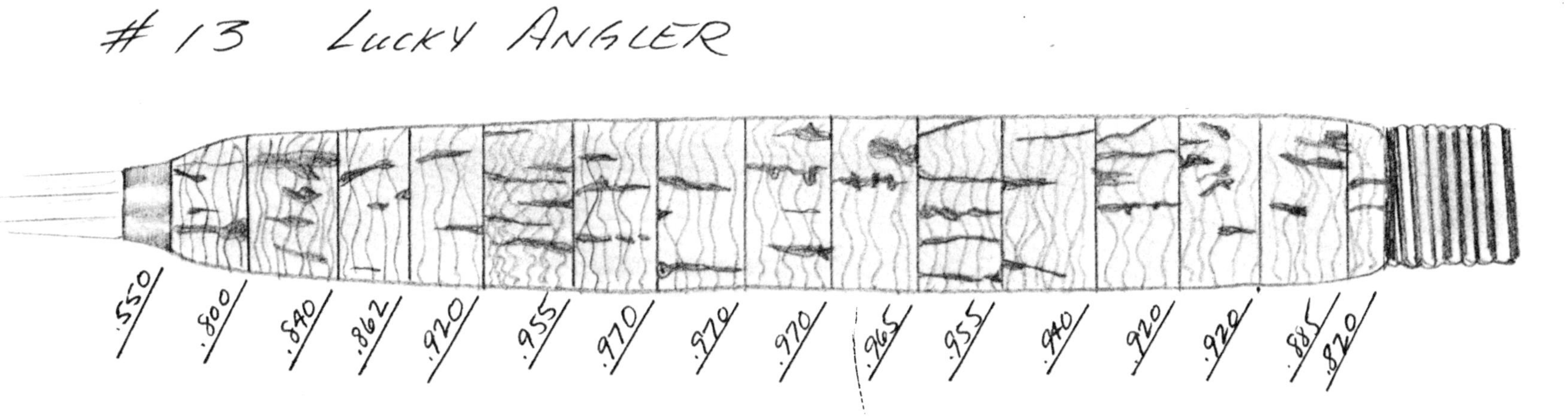

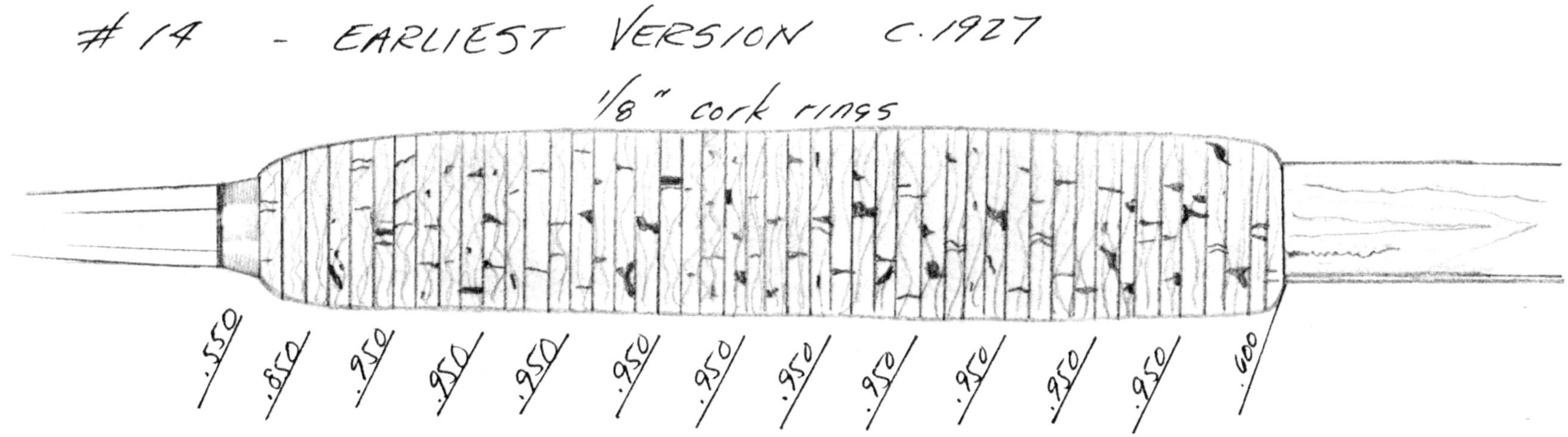

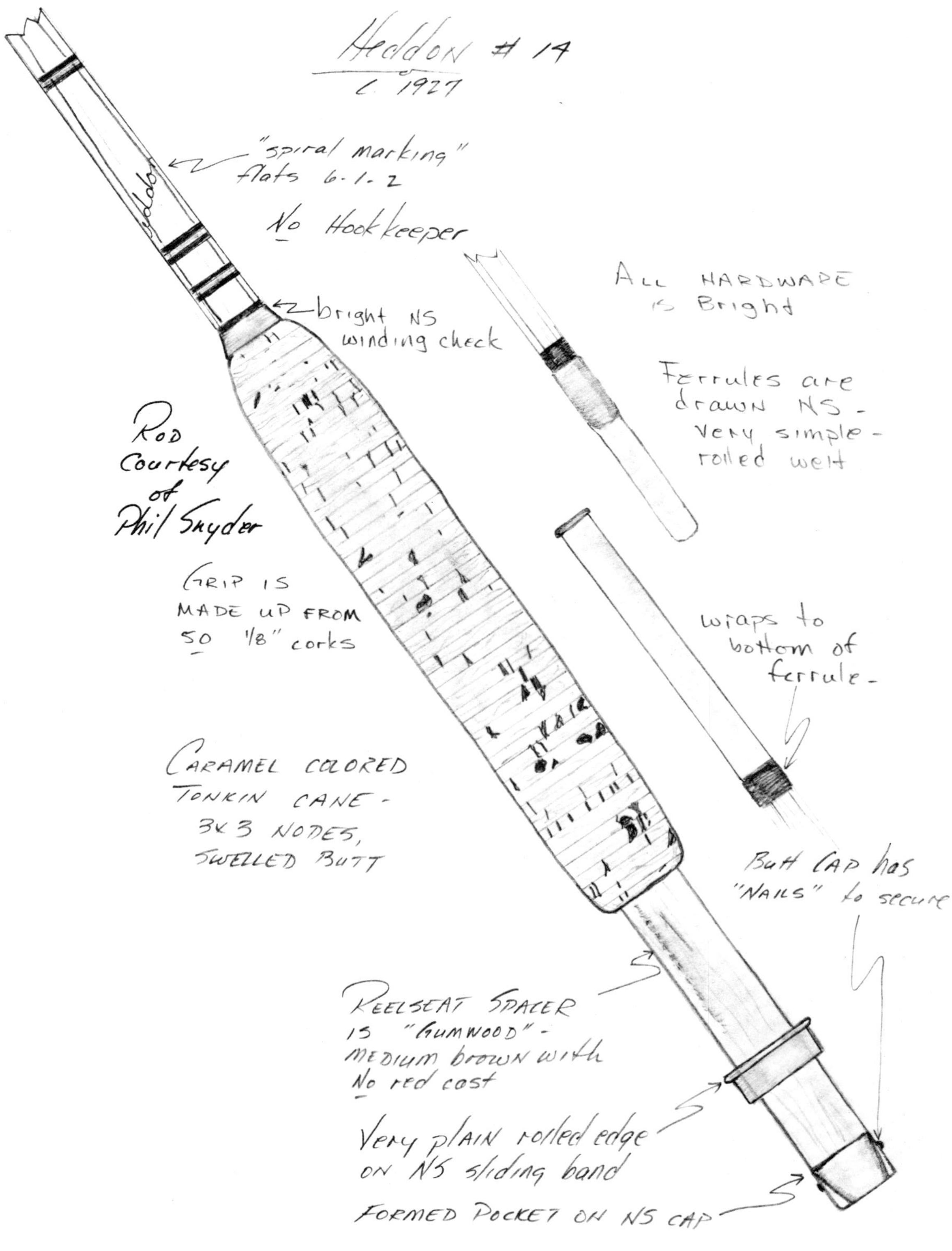

Heddon # 14
c 1927
"spiral marking"
flats 6-1-2
No Hookkeeper
bright NS
winding check
Rod
Courtesy
of
Phil Snyder
Grip is
made up from
50 1/8" corks
Caramel colored
Tonkin cane -
3x3 nodes,
swelled butt
All hardware
is Bright
Ferrules are
drawn NS -
very simple -
rolled welt
wraps to
bottom of
ferrule -
Butt cap has
"Nails" to secure
Reelseat spacer
is "Gumwood" -
medium brown with
No red cast
Very plain rolled edge
on NS sliding band
Formed pocket on NS cap

James Heddon's Sons — Hardware Specs — 1930 - 1933

Year Intro	Rod #	Model	Reelseat	Grip	Stripper	Snake	TT	Ferrules
1930	13	Lucky Angler	C&R Gumwood	Thick Cigar	PSS	PSS	PSS	Bright NS, RW
1930	14	Thorobred	C&R, NS	Bottle Half Wells	PC	PSS	PSS	Bright NS, RW
1931	17	Black Beauty	NS C&R, Cedar	Bottle Half Wells	A	PSS	PSS	NS. GM
	20	B.S. Favorite	Hold-Tite	Half Wells	A	PSS	PC	NS, CP
1930	21/31	Featherwt 2/2	BNS, C&R,	Cigar	A	PSS	PC	NS, CP
	35	Peerless	DL - Heddylin	Full Wells	A	PC	PC	NS, CP
	50	President	C&R, NS, Walnut	Rev Half Wells	A	PC	PC	BNS
	51	Princess	C&R, BNS, Cork	Cigar	A	PC	PC	BNS
1932	60	DeLuxe	Checkered Walnut	C&R Walnut	A	PC	PC	NS, GM, Pyralin
1930	120	Extralite #20	C&R, NS, CP	Teardrop	A	PSS	PC	NS, CP
1930	135	Extralite #35	C&R, BNS	Teardrop	A	PSS	PC	BNS
1930	150	Extralite #50	C&R, BNS	Teardrop	A	PC	PC	BNS

James Heddon's Sons — Wrap Specs — 1930 - 1933

Rod #	Model	Wrap Color	Tipping Color	Notes
13	Lucky Angler	Green #24 or Red #10	Red #10 or Green #24	Green tipped red is earlier
14	Thorobred	Red #10	Black #1	Mid-Butt: R, B, R
17	Black Beauty	Black #1	Orange #8	Simple signature at HK
20	B. S. Favorite	Maroon #12	Gold #5	Mid-Butt: G, M, G - has HK
21/31	Featherweight	Gold #7	Black #1	Has HK
35	Peerless	Gold #7	Black #1	Has HK
50	President	Gold #7	Green #15	Has HK
51	Princess	Grey #31	Black #1	Has HK
60	DeLuxe	Brown #27	Brown #28	Has HK
120	Extralite Stanley	Maroon #12	Gold #5	Mid-Butt: G, M, G - has HK
135	Extralite Peerless	Gold #7	Black #1	Has HK
150	Extralite President	Gold #7	Green #15	Has HK

James Heddon's Sons **Hardware Specs** **1934 - 1938**

Year Intro	Rod #	Model	Reelseat	Grip	Stripper	Snake	TT	Ferrules
1934	10	Blue Waters	NS, SB	Straight	PS	PS	PS	NS, RW
1930	13	Lucky Angler	Any Heddon style	Thick Cigar	PS	PS	PS	NS, RW
	14	Thorobred	UL, Cedar/Pyralin	Half Wells	Cy	PS	TS	NS, GM, HW, S
1931	17	Black Beauty	UL, Cedar\Pyralin	Half Wells	A	PS	PS	NS, GM, HW, S
1931	19	Riptide	NS, SB, Chromed	Half Wells	PC	PC	PC	NS, CP
	20	B.S. Favorite	UL, Cedar/Pyralin	Half Wells	A	PS	PS	NS, GM, HW, S
1930-33	31	Featherwt. 2/2	C&R, NS, CP	Cigar	A	PS	PS	NS, C
	35	Peerless	C&R, NS	Rev Half Wells	A	PS	PS	NS, GM, HW, S
	50	President	NS, SB or Locking	Rev Half Wells	A	PS	PS	BNS, HW, S
	51	Princess 2/2	BNS, C&R, Cork	Cigar	A	PS	PS	BNS, HW, S
	60	DeLuxe	C&R, BNS w/Half Wells Walnut grip		A	PS	PS	NS, GM, HW, S
	1000							

Rod #	Model	Wrap Color	Tipping Color	Notes
10	Blue Waters - 1934	Blue #17	White #2	Exposed W/C, Open HK
10	Blue Waters - 1935	Maroon #12	Green #24	Exposed W/C, Open HK
13	Lucky Angler	Orange #8	Black #1	Exposed W/C, Open HK
14	Thorobred	Red #10	Black #1	Without W/C, Open HK
17	Black Beauty	Black #1	Orange #8	Without W/C, Open HK
20	Bill Stanley's Favorite	Maroon #12	Gold #5	Without W/C, Open HK
35	Peerless	Brown #27	Dark Brown #30	Without W/C, Open HK
50	President	Brown #27	Dark Brown #30	Without W/C, Open HK
60	DeLuxe	Brown #27	Dark Brown #30	Without W/C, Open HK

James Heddon's Sons — Hardware Specs — 1939 - 1947

Rod #	Model	Reelseat	Grip	Stripper	Snake	TT	Ferrules
10	Blue Waters	DL Pyralin	Thick Cigar	Ag	PSS	PSS	NS, RW
13	Lucky Angler	DL Pyralin	Cigar	PSS	PSS	PSS	NS, RW
14	Thorobred	DL, Cedar/Pyralin	Heddon Half Wells	PSS	PSS	PSS	NS, GM, HW, S
17	Black Beauty	DL, Tenite	Heddon Half Wells	PSS	PSS	PSS	NS, GM, HW, S
20	B. S. Favorite	DL, Pyralin	Heddon Half Wells	A	PSS	PSS	NS, GM, HW, S
35	Peerless	DL, Walnut/Pyralin	Rev Half Wells	A	PSS	PSS	NS, GM, HW, S
50	President	Walnut/Pyralin SDL	Rev Half Wells	A	PSS	PSS	NS, GM, HW, S
60	Deluxe	BNS, C&R	Walnut Wood Grip	A	PSS	PSS	NS, GM, HW, S
1000	Rod of Rods	Gold, Walnut/Pyralin	Straight	A	Gold	Gold	NS, GM, HW, S
	Featherweights (Available in all Model Numbers #14 and Higher)	BNS, C&R or SDL	Teardrop	A	PSS	PSS	NS, GM, HW, S

James Heddon's Sons — Wrap Specs — 1939 - 1947

Rod #	Model	Wrap Color	Tipping Color	Notes
10	Blue Waters	B/Maroon Jasper	None	Exposed W/C, Open HK
13	Lucky Angler - 1940	Orange #8	Black #1	Exposed W/C, Open HK
13	Lucky Angler - 1941-43	Grey #31	Black #1	Exposed W/C, Open HK
14	Thorobred	Red #10	N/A	W/O W/C, Open HK
17	Black Beauty	Black #1	Orange #8	W/O W/C, Open HK
20	B. Stanley's Favorite	Maroon #12	Gold #5	W/O W/C, Open HK
35	Peerless - 1941+	Brown #27	Black #1	W/O W/C, Open HK
50	President - 1941+	Brown #27	Green #24	W/O W/C, Open HK
60	DeLuxe	Brown #27	Dark Brown #30	W/O W/C, Open HK
1000	Rod of Rods	Brown #27	Dark Brown #30	1939 Only
1000	Rod of Rods	Dark Brown #30	Gold #7	Custom Wraps Invited

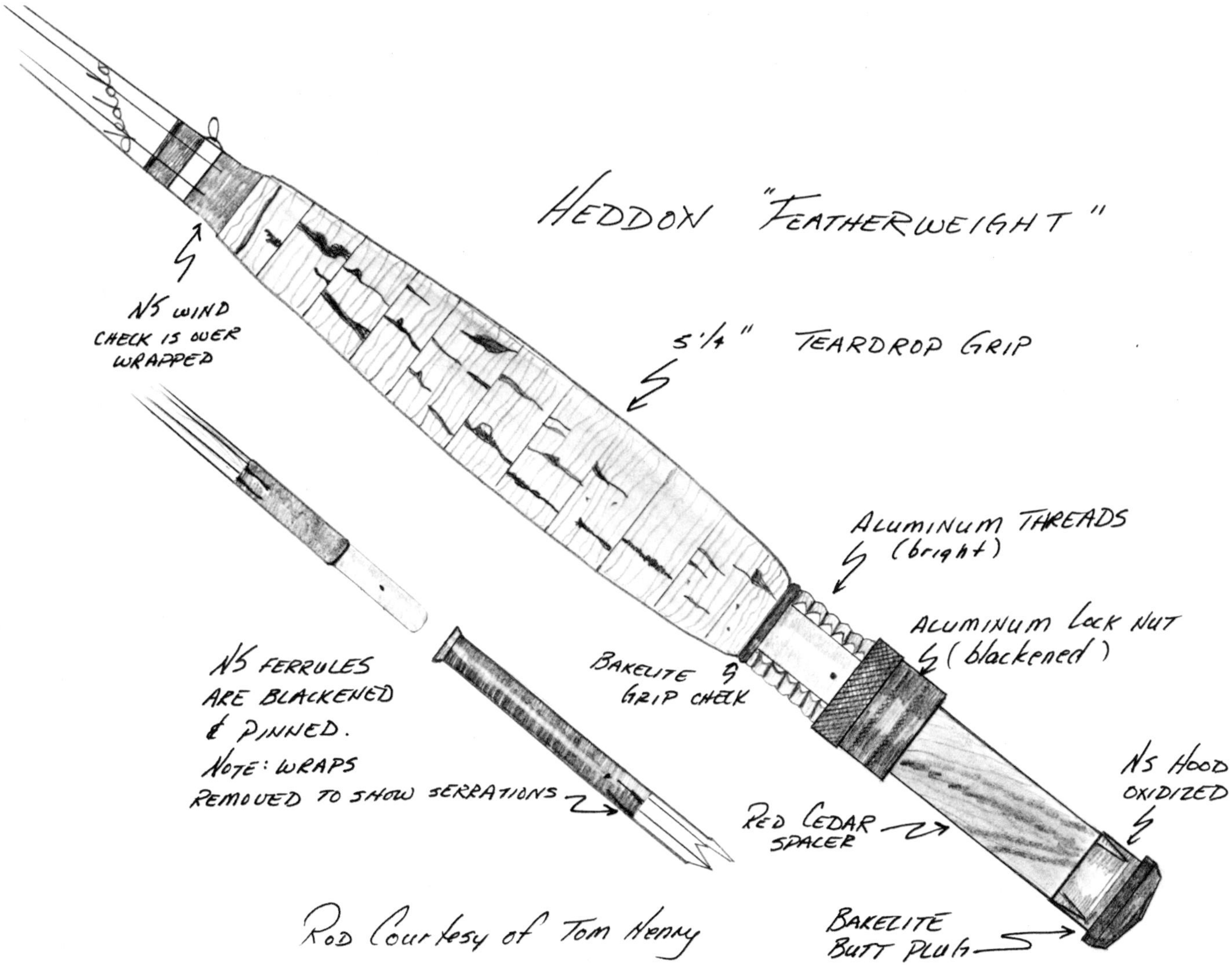

HEDDON "FEATHERWEIGHT"
5 1/4" TEARDROP GRIP
NS WIND CHECK IS OVER WRAPPED
ALUMINUM THREADS (bright)
ALUMINUM LOCK NUT (blackened)
BAKELITE GRIP CHECK
NS FERRULES ARE BLACKENED & PINNED.
NOTE: WRAPS REMOVED TO SHOW SERRATIONS
RED CEDAR SPACER
NS HOOD OXIDIZED
BAKELITE BUTT PLUG
Rod Courtesy of Tom Henry

James Heddon's Sons — Hardware Specs — 1948 - 1954

Rod #	Model	Reelseat	Grip	Stripper	Snake	TT	Ferrules
8	Heddon ('51/52)	SDL - Brown	Straight	PSS	PSS	PSS	NS, RW
10	Blue Waters	SDL - Pyralin	Thick Cigar	PSS	PSS	PSS	NS, RW
14	Thorobred	SDL - Coral	Half Wells	PC	PSS	PSS	NS, GM, HW, S
17	Black Beauty	SDL - Black	Half Wells	PC	PSS	PSS	NS, GM, HW, S
20	Bill Stanley	SDL - Brown	Half Wells	PC	PSS	PSS	NS, GM, HW, S
35	Peerless	SDL, Wood/Pryalin	Rev Half Wells	PC	PSS	PSS	NS, GM, HW, S
50	President	SDL, Walnut/Pyralin	Rev Half Wells	PC	PSS	PSS	NS, GM, HW, S
1000	Rod of Rods	SDL, Walnut/Pyralin	Straight	A	Gold	Gold	NS, GM, HW, S
125	Expert	SDL - Brown	Straight	Gold	Gold	Gold	Gold NS, HW S
115	Premier	SDL – Brown	Straight	Gold	Gold	Gold	Gold NS HW, S

James Heddon's Sons — Wrap Specs — 1948 - 1954

Rod #	Model	Wrap Color	Tipping/Trim Color	Notes
8	Heddon Fly Rod	Rainbow Varigated	None	Exposed W/C, No HK
10	Blue Waters	Maroon/Black Jasper	None	Exposed W/C, Open HK
14	Thorobred	Red #10	None	w/o W/C, Open HK
17	Black Beauty	Black #1	Orange #8 at HK	w/o W/C, Open HK
20	Bill Stanley	Maroon #12	Gold #5	w/o W/C, Open HK
35	Peerless	Old Gold #7	Black #1	w/o W/C, Open HK
50	President	Gold #7	Green #23	w/o W/C, Open HK
1000	Princess	Gold #7	Brown #38	Custom Wrpa Colors Invited
125	Expert	Burgundy #12	Gold #5	w/o W/C, Open HK
115	Premier	Burgundy #12	Gold #5	w/o W/C, Open HK

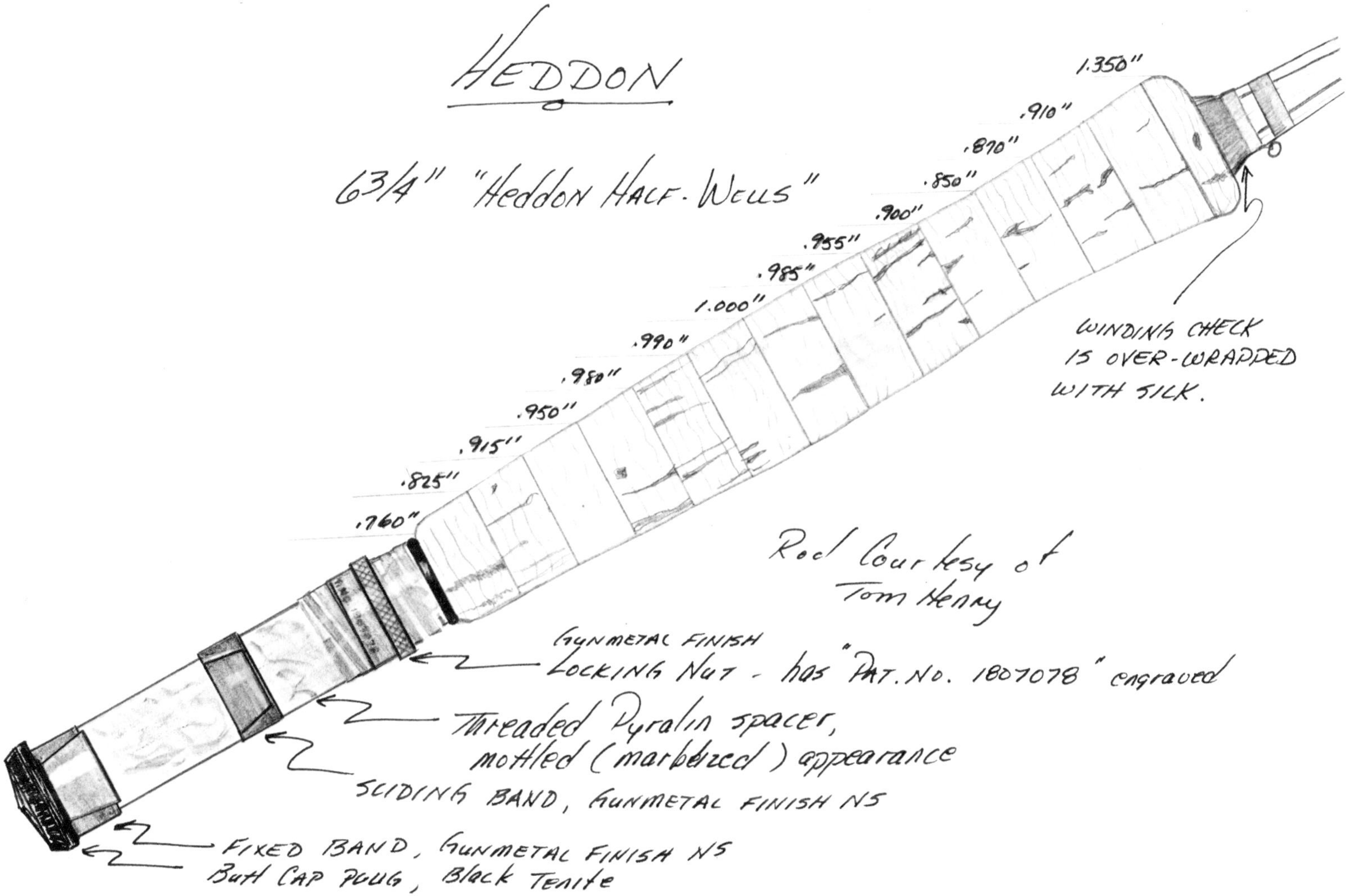

HEDDON
6 3/4" "Heddon Half-Wells"
1.350"
.910"
.870"
.850"
.900"
.955"
.985"
1.000"
.990"
.980"
.950"
.915"
.825"
.760"
WINDING CHECK IS OVER-WRAPPED WITH SILK.
Rod Courtesy of Tom Henry
GUNMETAL FINISH
LOCKING NUT - has "PAT. NO. 1807078" engraved
Threaded Pyralin spacer, mottled (marbelized) appearance
SLIDING BAND, GUNMETAL FINISH NS
FIXED BAND, GUNMETAL FINISH NS
Butt Cap PLUG, Black Tenite

The Horrocks-Ibbotson Company

The World's Largest Manufacturer of Fishing Tackle had a very small beginning. The company goes back to 1812 when John Osborn opened a small factory to manufacture buckskin and kid gloves. Some years later, this operation was taken over by Chester D. Clark. In 1854, his son, George A. Clark, took over the business.

George Clark's brother was accidentally killed in a Chicago fire in 1857. At this time, the country was in the grips of an economic Panic. George Clark had endorsed quite heavily for his brother and had to make good on the endorsements. This he did cheerfully to sustain the honor of his own and his brother's name, although he in no way benefited by the endorsements. It took him seven years to pay the notes, principal and interest. He was a young man and had but just started his own business and this loss was a great set-back to him. Nevertheless he did not regard it in that light. The fact was known to businessmen generally and it established his reputation as a man of honor and integrity among those of the world of trade who had not before known him.

In 1863 Clark hired an English immigrant as a clerk. His name was Hugh James Horrocks. The 23-year-old Jim Horrocks was an avid angler, and convinced George Clark to import some English tackle for sale in the store. The trial proved so successful that in 1880 Jim Horrocks and a salesman named James England were made partners in the firm which was called George A. Clark & Co.

When Mr. England died in 1891, the company name was changed to The Clark-Horrocks Company. The primary business of the firm then became the manufacture of fishing tackle of all sorts. In 1894, a new errand boy was hired; Edward Dunscomb Ibbotson, a recent graduate of the Eastman Business School of Poughkeepsie. Ibbotson was a natural salesman, and was also an enthusiastic angler. When the firm's name was changed to Clark-Horrocks Co., Ed Ibbotson became a director.

Clark-Horrocks Co. bought out the Trenton, N.J. firm of George Kempf in 1900. They next acquired the John B. McHarg Co. of Rome, N.Y. in 1901. That firm had been engaged in the manufacturing of fishing flies. In 1905 the company built a new factory on Whitesboro Street in Utica to supply their retail outlet at 54 Genesee Street. Both operations continued to grow until they were widely known throughout the world. George A. Clark died in 1908. After his death, the firm was re-incorporated in 1909 as the Horrocks-Ibbotson company. H. James Horrocks was President, and Edward D. Ibbotson was the Vice-President. Horrocks-Ibbotson was able to acquire several other tackle companies in this era.

Abbey & Imbrie (*Fishing Tackle Fit For Fishing*), Y&E Manufacturing Company (maker of Yawman & Erbe automatic reels), American-Swiss Magneto Company (makers of fishing reels) Winchester Repeating Arms Co. (reel and bait, steel and bamboo rod division), G.H. Mansfield Co. (reel manufacturing division), Wm. Devine (floatmaker), the Bait Manufacturing Division of Gillette Bros. Co. and the H. J. Frost Co. (makers of bamboo rods) were purchased and incorporated into H-I.

Jim Horrocks became ill in 1926, and died in 1927. Edward Ibbotson became the President of the company and the Horrocks interest in the business passed to his grandson, Richard H. Balch. Mr. Balch became President of the company in 1942. Ed Ibbotson was Chairman of the Board until his death in 1958. In 1962, Horrocks-Ibbotson celebrated its 150th anniversary. Visitors to the H-I plant were greeted by a large oak and glass showcase - one of the original cases from the James Osborn factory store. Inside the case were samples of the product responsible for the start of the World's Largest Manufacturer of Fishing Tackle. The case displayed buckskin and kid gloves.

The rods made by Horrocks-Ibbotson in the years up to 1935 filled every niche in the rodmaking business. The high grade rods such as the President and higher were as good as any rods being produced by the competition. On the other end of the spectrum were the genuine production rods (the lowest grade #08311 sold at retail for 85¢, and was *put up in a paper bag*). There is no need to discuss the quality level of this model!

Perhaps it is best to say that H-I made hundreds of different models throughout the years. It would not be possible to catalog all of them today. The charts in this chapter demonstrate the variety of models, but these are only a few of those listed in the catalogs. There are 50 different numbered rod models that had such minute differences in fittings and wraps as to make them nearly indistinguishable without side-by-side comparison. Listed in this book are most of the named rods, especially those that will be seen most often.

The decals shown on the flat map are somewhat useful for dating rods. The Clark-Horrocks & Co. decal has been seen only once - it is quite rare and dates from c. 1900. The diamond with the UTK logo is from c. 1905 to WW I (1919). This UTK Diamond logo is usually seen stamped into the reelseat, but does appear in decal form. The Trout Logo decal was used from c. 1923 to c. 1929. It is rarely seen and is easily the most beautiful of the decals. The trout is full color, complete with pink stripe, the sun is gold, and sets over a golden sea.

Next to appear was the elongated Double Diamond with "Utica, NY" inside. In its early form, only the Diamond was used - the banner reading *Best By Test* was not used on rod decals until later, though it was seen in catalogs and letterheads dating from 1929. The outer (larger) Diamond is gold foil with black border and printing. The inner (small) diamond is red with black print. This version was used from c. 1929 to c. 1933. In 1934, the entire double-diamond logo including the banner was used. In this form, the banner is royal blue with white letters, the outer diamond is red and the inner diamond is gold foil. This decal was used from c. 1934 to c. 1939.

The fanciest of all H-I decals was the one used from c. 1945 to c. 1950. This logo featured a bright red "H-I" on a white diamond. The diamond is accompanied by two banners. The upper banner is yellow with "Fish Rod" in black. The lower banner is turquoise with black print. Sometimes the lower banner includes the words Utica, N.Y. and Made in U.S.A. The words *Genuine Tonkin Cane* are turquoise edged with black. The decal of the early 1950's was rectangular. This was a subdued design, with black print on an olive green background. The small diamond logo is gold foil.

The last H-I decal is a simple red diamond with a large white H-I. This logo was used beginning in 1951 in print, but did not appear on bamboo rods except those that were factory refinishes in the 1950's. The location of the decal was sometimes higher on the rodshaft than is shown on the flat map. This is especially true of rods with elaborate signature wraps and those with intermediate wraps.

If you are trying to identify a rod that has no decal, but you think it is an H-I product, look at the writing on the rodshaft. H-I used white ink, and usually wrote with the words running toward the grip. The only other maker that used white ink was Edwards, who usually wrote with the words reading away from the grip. You should have no difficulty in determining whether you are looking at an H-I vs. an Edwards. The quality difference is obvious even to a tyro.

As with all rods, the most recent H-I products are the most commonly seen. The rods from 1940 through 1953 look pretty much the same. The reelseat style did not change much, and the spacers were usually solid color plastics before the war, and marbleized plastics after the war. The cuts show the entire 1949 H-I rod line; note that all rods had similar reelseats, and the grips are shaped just enough to discern a style. These rods were just about as plain-vanilla as you could ever find, but had a tremendously loyal following.

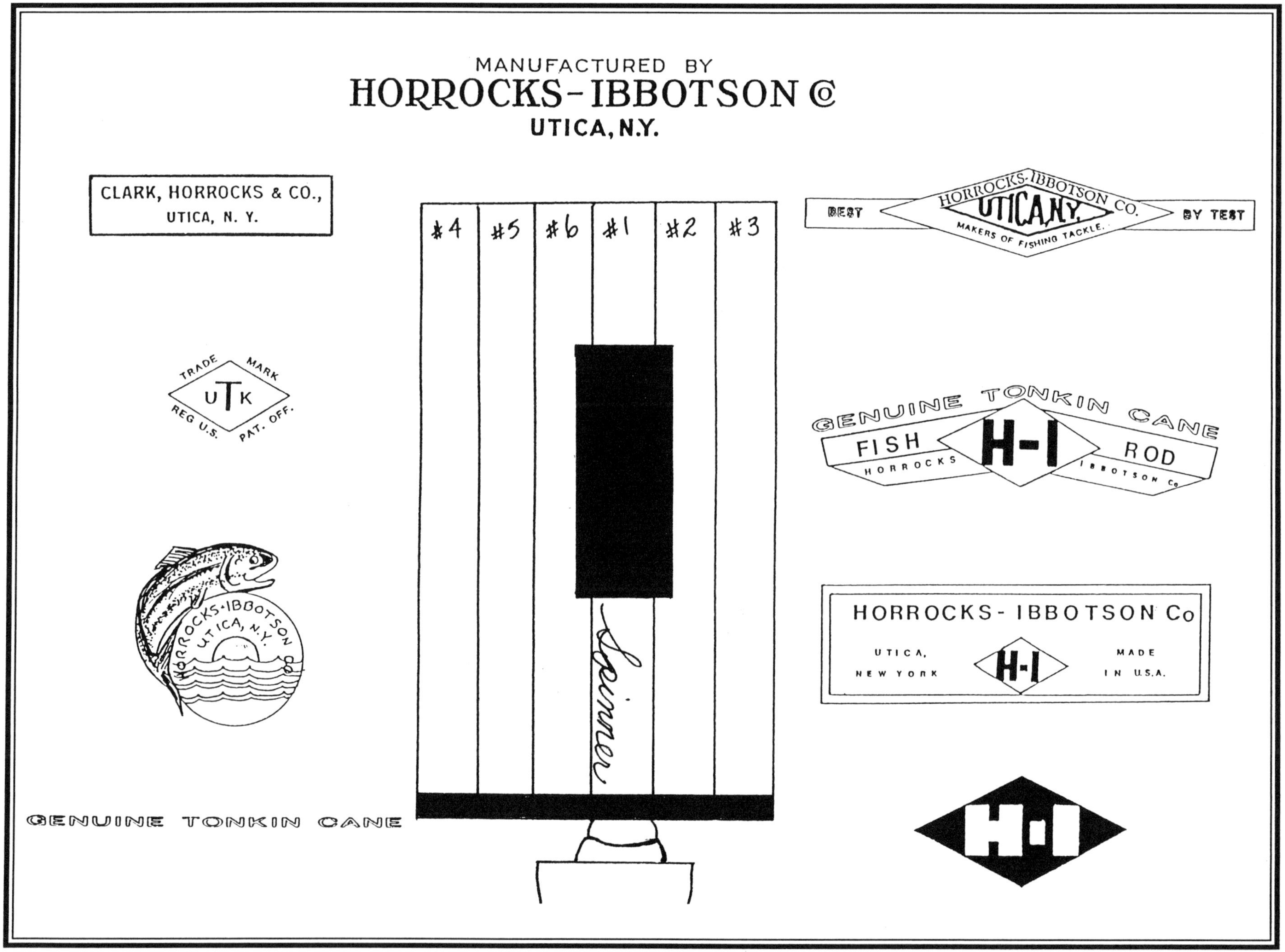
MANUFACTURED BY
HORROCKS-IBBOTSON Co
UTICA, N.Y.
CLARK, HORROCKS & CO.,
UTICA, N. Y.
TRADE MARK
UTK
REG U.S. PAT. OFF.
HORROCKS-IBBOTSON Co
UTICA, N.Y.
GENUINE TONKIN CANE
#4
#5
#6
#1
#2
#3
Spinner
BEST
HORROCKS-IBBOTSON CO.
UTICA, N.Y.
MAKERS OF FISHING TACKLE.
BY TEST
GENUINE TONKIN CANE
FISH
H-I
ROD
HORROCKS
IBBOTSON Co
HORROCKS-IBBOTSON Co
UTICA,
NEW YORK
H-I
MADE
IN U.S.A.
H-I

Horrocks–Ibbotson Co. Hardware Specs Early 1920's

Rod #/Name	$	Reelseat	Grip	Stripper	Snake	TT	Ferrules
President	18.00	NS, SB	6" Cigar	A	HS	HS	NS, HW, S reinforced
Mohawk	15.00	NS, SB	6" Cigar	HS	HS	HS	NS, HW, S, reinforced
2380 Tournament	10.00	NS, SB	5½" Cigar	HS	E, HS	HS	NS, HW, S, reinforced
2386 Sportsman's	10.00	NS, SB	5½" Cigar	HS	HS	HS	HS, HW, S, reinforced
2384 Defiance	10.00	NS, SB-wood	5½" Cigar	HS	HS	HS	HS, NW, S, reinforced
Expert	8.00	NS, SB-cedar	5½" Cigar	HS	HS	HS	HS, NW, S, reinforced
UXLD	7.50	NS, SB	5½" Cigar	NS	NS	NS	NS, HW, S, reinforced
Senator	6.00	NPB, C&R	5" Cigar	Ring	Ring	NS	NS, HW, S
2354 African Steel	6.00	NPB, SB	5" sheet	Snake	HS	HS	NPB, HW
2353 African Steel	5.00	NPB, SB	5" sheet	Tie	Tie	NS	NPB, HW
2334 Togo Special	5.00	NPB, SB	5" sheet	NS	NS	NS	NPB, HW
2342 Black Beauty	5.00	BNPB, SB	5" Celluloid	E, HS	E, HS	HS	BNPB, HW
2336	4.00	NPB, SB	5" sheet	HS	HS	HS	NPB, HW, reinforced
2315 Steel Cane	2.00	NPB, SB	5" sheet	NS	NS	NS	NPB, RW
2313 Hexrod	1.50	NPB, C&R	4½" sheet	S	S	S	NPB, RW
2311	1.00	NPB, C&R	4½" wood	Tie	Tie	S	NPB
02311	.85	Same as #2311, with no extra tip section					

Horrocks–Ibbotson Co. Wrap Specs Early 1920's

Rod #/Name	Main Color	Trim #1	Trim #2	Notes
President	B&W Jasper	Orange #8	Black #1	Triple cluster winds, B-O-B, O-B-O, etc
Mohawk	Red #10	Yellow #3	Green #24	Double Tipping, triple clusters
2380 Tournament	B&W Jasper	Yellow #3	None	
2386 Sportsman's	Red #10	Yellow #3	None	Close intermediates, yellow
2384 Defiance	Green #24	Yellow #3	Red #10	Tipped Yellow & red; intermediates yellow
Expert	Black #1	Red #10	None	Close intermediates, red
UXLD	Red #10	Black #1	Orange #6	Triple cluster winds, B-O-B, O-B-O, etc
Senator	Red #10	Black #1	None	Close intermediates, red
2354 African Steel	Red #10	Green #24	None	Broad green bands on Butt & mid
2353 African Steel	Green #24	Yellow #3	None	Triple cluster as intermediates, Y-Y-Y
2334 Togo Special	Red #10	Black #1	None	Triple cluster as intermediates, B-R-B, R-B-R
2342 Black Beauty	Black #1	None	None	Very close intermediates, black
2336	Red #10	Brown #28	Black #1	Brown border wraps, black intermediates
2315 Steel Cane	Purple #16	None	None	Very close intermediate
2313 Hexrod	Red #10	Black #1	None	Wide intermediates, alternate red & black
2311	Black #1	None	None	Wide intermediates, black
02311	Yellow #3	None	None	Wide intermediates, black

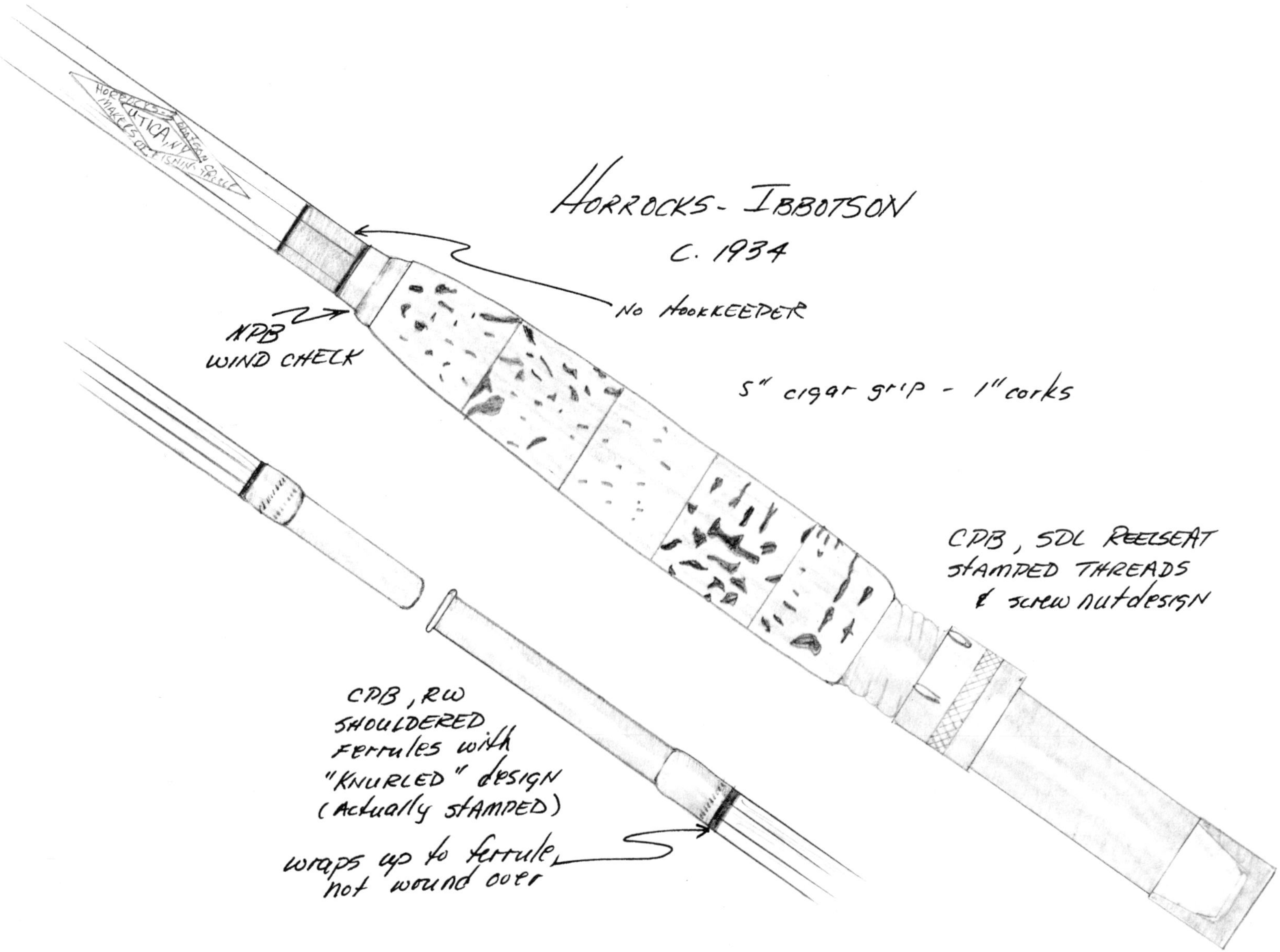

HORROCKS - IBBOTSON
C. 1934
NO HOOKKEEPER
NPB
WIND CHECK
5" cigar grip - 1" corks
CPB, SDL REELSEAT
stAMPED THREADS
& screw nut design
CPB, RW
SHOULDERED
Ferrules with
"KNURLED" design
(Actually stAMPED)
wraps up to ferrule,
not wound over

Horrocks–Ibbotson Co. Hardware Specs 1930's

Rod Name	$	Reelseat	Grip	Stripper	Snake	TT	Ferrules
Cunningham	30.00	Bronze Reeltite	6½" mild HW	A	CS	A/CS	NS, HW, S reinforced
Vernley	20.00	NS Reeltite	Peckerhead	A	CS	A/CS	NS, HW, S, reinforced
Jay-Dee-Jay	20.00	NS Reeltite	6" Cokebottle	CS	CS	CS	NS, HW, S, reinforced
Willquin	20.00	NS, C&R	7" Cigar	A	CS	A/CS	HS, HW, S, reinforced
UNXLD	16.50	NS Reeltite	6" Sturgin	A	CS	A/CS	HS, NW, S, reinforced
President	16.50	NS, Scrutite	7" Rev. HW	CS	CS	CS	HS, NW, S, reinforced
Featherweight	16.50	NS, C&R-cork	5½" Full Wells	A	CPNS	A	NS, HW, S, reinforced
Golden Rod	14.00	CPNS Reeltite	Walnut	A	CS	A/CS	NS, HW, S
Davis	14.00	CPNS, C&R	6" Straight	A	CS	A/CS	NPB, HW
Harvey	14.00	C&R-cork	6" Rev. HW	CS	CS	CS	NPB, HW
The Old Gold	14.00	CPNS, C&R	6" Rev. HW	CS	CS	CS	NPB, HW
Spinner	16.00	CPNS Reeltite	6" Straight	CS	CS	CS	BNPB, HW
Favorite	11.00	NS, C&R	6" Rev. HW	A	CS	A/CS	NPB, HW, reinforced
Governor	12.00	NS, C&R	6" Sturgis	A	CS	A/NS	NPB, RW
Senate (round)	12.00	NPB, C&R	6" Cigar	A	CS	CS	NPB, RW
Fairy	10.00	NPB, C&R	6" Rev. HW	Snake	CS	CS	NPB
Lucky Strike	5.00	NPB, C&R	5½ Full Wells	Ag	CS	A/CS	NPB, RW
Ike Walton	10.00	NPB Reeltite	6" Cigar	CS	CS	CS	NPB, HW
Adirondack	7.50	NPB, C&R	6" Cigar	CS	CS	CS	NPB, HW
Catskill	8.50	NPB, C&R	6" Cigar	CS	CS	CS	NPB, RW
Oriskany	7.50	CPNS Reeltite	Walnut	CS	CS	CS	CPNS, HW, S
Nat'l Sportsman	7.00	NPB, C&R	6" Sturgis	CS	CS	CS	NPB, RW

Continued Next Page

Horrocks–Ibbotson Co. Hardware Specs 1930's

Rod Name	$	Reelseat	Grip	Stripper	Snake	TT	Ferrules
Commodore	6.00	CPB, C&R	Velgreen	CS	CS	CS	NPB, RW
Beaverkill	5.75	CPB, C&R	5½" Straight	CS	CS	CS	CPB, RW
Utica	5.00	CPB, C&R	5½" Cigar	Ag	CS	A/CS	CPB, RW
Roosevelt	4.25	NPB, C&R	5½" Fat Cigar	CS	CS	CS	CPB, RW
Mayflower	4.00	NPB, C&R	5½" Cigar	CS	CS	CS	CPB, RW
Hudson	4.00	NPB, C&R	5½" Cigar	CS	CS	CS	CPB, RW
Columbia	3.75	NPB, C&R	5½" Cigar	CS	CS	CS	CPB, RW
Flash (1 tip)	3.00	CPB, C&R	5½" Cigar	CS	CS	CS	CPB, RW
Majestic	2.25	NPB, C&R	Sheet Cork	CS	CS	CS	NPB, RW
02311	1.75	NPB, C&R	Composite	CS	CS	CS	NPB, No welt
Two-piece Rods							
Princess	7.50	CP, C&R	5½" Cigar	CS	CS	CS	CPB, RW
Queen	14.00	CPB, C&R	6½" Cigar	A	CS	CS	CPNS, HW, S
King	18.00	CPNS Scrutite	7" Cigar	A	CS	A/CS	CPNS, HW, S

Horrocks–Ibbotson Co. Wrap Specs 1930's

Rod Name	Main Color	Trim #1	Trim #2	Notes
Cunningham	Gold #7	Green #24		Triple border wraps, green
Vernley				Close intermediates
Defiance	B&W Jasper #X	Purple #16		
Willquin	Orange #8	Green #24		Cluster of 6 orange, 1 green intermediate
UNXLD	Green #24	White #2		Clusters G-W-G-W-G
President	Red #10	White #2		Wide intermediates, red
Featherweight	Black #1	Orange #8	White #2	Tipped white & orange, intermediates orange
Golden Rod	Gold #5	None	None	Close intermediates, gold
Davis	Yellow #3	Black #1		No ornamental wraps
Harvey	B&W Jasper #X	None		No ornamental wraps
The Old Gold	Red #10	Green #24		Regular intermediates, red
Spinner	Purple #16	Gold #7		No ornamental wraps
Favorite	Red #10	White #2		Triple clusters, R-W, R
Governor	B&W Jasper #X	Red #10	Yellow #3	Border wraps yellow, Clusters Y-R-Y
Senate (round)	Green #24	Gold #5		Broad green bands on butt & mid, intermediates gold
Fairy	Green #24	Red #10		Wide spaced intermediates, green
Lucky Strike	Red #20	Yellow #3		Wide spaced intermediates, yellow
Ike Walton	Red #20	Green #24		Clusters: R-R-R-R, G-G-G, R-R-R-R
Adirondak	Purple #16	None		Closed intermediates- double, single, double, etc.
Catskill	Red #10	Black #1	White #2	Tipped B-W-B, intermediates black
Oriskany	Black #1	Gold #7		Close ingermediates, gold
Nat'l Sportsman	Red #10	Green #24		Close intermediates, green

Continued Next Page

Horrocks–Ibbotson Co. Wrap Specs 1930's

Rod Name	Main Color	Trim #1	Trim #2	Notes
Comodore	Red #10	Green #24		No ornamental wraps
Beaverkill	Gold #7	Red #10	Black #1	Gold, tipped red/black, intermediates red
Utica	Orange #8	Black #1		Wide spaced intermediates, orange
Roosevelt	Red #10	Gold #7		Very wide red intermediates tipped gold
Mayflower	Orange #8	None		No ornamental wraps
Hudson	Red #10	White #2		Red intermediates
Columbia	Black #1	None		Wide spaced intermediates, black
Flash	Lime #19	None		No ornamental wraps
Majestic	Red #10	Green #24		Wide spaced intermediates, green
02311	Green #24	None		Very plain
Two-Piece Rods:				
Princess	Blue #17	White #2	None	Regular tipping only
Queen	Green #24	White #2	None	Regular tipping only
King	Red #10	White #2	None	Regular tipping only

Horrocks–Ibbotson Co. Hardware Specs 1940's

Rod Name	$	Reelseat	Grip	Stripper	Snake	TT	Ferrules
Cunningham	36.00	SDL, NS	7" Mild HW	A	PC	PC	CPNS, HW, S, w/o
HI-Test	19.12	SDL, NS/Bakelite	7" Half Wells	A	CS	PC	NS, HW, S, w/o
Adirondack	9.37	SDL, CPB/Bakelite	6½" Cigar	Ag	CS	CS	NS, RW, w/o
#FF Beaverkill	7.62	SDL, CPB	6½" Straight	CS	CS	CS	CPNS, RW, w/o
2324 Lucky Strike	6.65	SDL, CPB	6½" Full Wells	Ag	CS	Ag/CS	CPB, RW, SH
2305 Roosevelt	5.12	SDL, CPB	6" Cigar	CS	CS	CS	CPB, RW, SH
Princess 2/2	8.25	SDL, NS/Bakelite	6" Cigar	Ag	CS	CS	NS, HW, S w/o
Golden Rod	14.25	SDL, NS/Bakelite	6½" Rev. HW	A	CS	Ag/CS	NS, HW, S, w/o

Horrocks–Ibbotson Co. Hardware Specs 1950's

Rod Name	$	Reelseat	Grip	Stripper	Snake	TT	Ferrules
Old Hi's HI-Test	50.00	SDL, CPB, Tenite	6½" Straight	PC	PC	PC	NS, HW, S, w/o
Chancellor	40.00	SDL, Black, Tenite	6½" Straight	PC	PC	PC	NS, HW, S, w/o
Vernley	30.00	SDL, AL, Tenite	6½" Full Wells	SS	TS	TS	NS, HW, S, w/o
Canada Creek	25.00	SDL, AL, Tenite	6½" Straight	SS	TS	TS	NS, HW
Cascade	20.00	SDL, CPB, Tenite	6½" Straight	SS	TS	TS	NS, HW
Spinner	15.00	SDL, AL, Tenite	6½" Half Wells	TS	TS	TS	NPB, RW, SH
Governor	12.00	SDL, AL, Tenite	6½" Rev. HW	TS	TS	TS	NPB, RW, SH
Beaverkill	9.00	SDL, AL, Tenite	6½" Straight	TS	TS	TS	NPB, RW, SH
Two-piece Rods							
Tonka Queen	15.00	SDL, Green, Tenite	6" Straight	PC	TS	TS	CPB, HW
Tonka Prince	11.00	SDL, AL, Tenite	6" Straight	TS	TS	TS	NPB, RW, SH

Horrocks–Ibbotson Co. Wrap Specs 1940's - 50's

Rod Name	Main Wraps	Tipping	Notes
Cunningham	Gold #6	Green #23	Impregnated, Steel core
HI-Test	Gold #6	Maroon #13	Impregnated
Chancellor	Yellow #3	Black #1	Impregnated
Vernley	Red #10	White #2	Impregnated
Canada Creek	Brown #28	Yellow #3	Impregnated
Cascade	Green #24	Black #1	Impregnated
Spinner	Purple #16	Yellow #3	Hexi-Super cane
Golden Rod	Gold #6	None	Hexi-Super cane
Governor	Red #10	None	Flame-finish cane
Adirondack	Purple #16	None	Medium brown cane
Beaverkill	Red #10	None	Flame-finish cane
Lucky Strike	Green #23	Nonw	Medium brown cane
Roosevlet	Brown #28	None	Medium brown cane
Two-Piece Rods:			
Princess	Blue #17	White #2	Medium brown cane
Tonka Queen	Gold #6	Brown #28	Browntone cane
Tonka Prince	Blue #17	None	Flame-finish cane

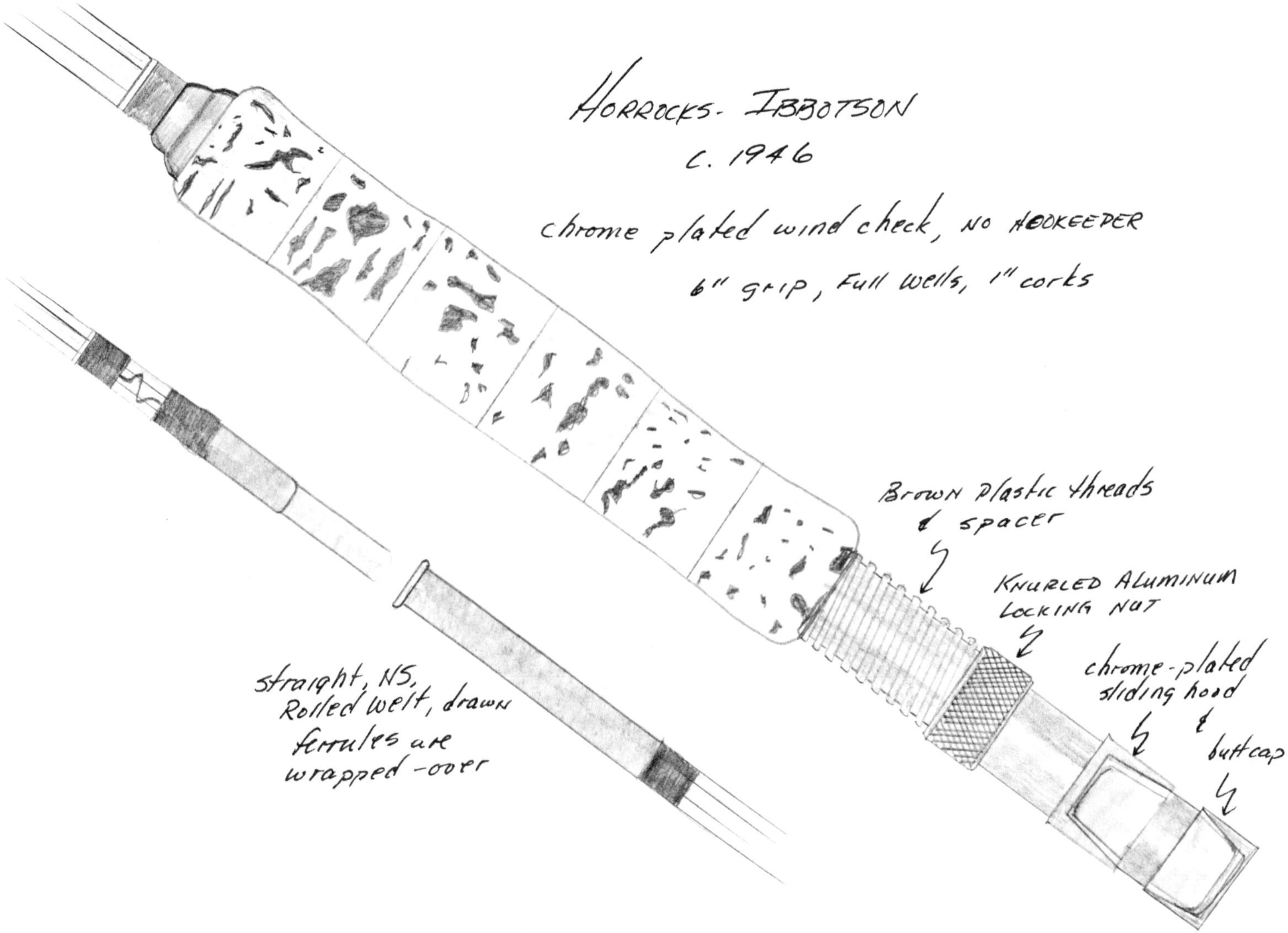
HORROCKS- IBBOTSON
c. 1946
chrome plated wind check, NO HOOKEEPER
6" grip, Full wells, 1" corks
Brown plastic threads
& spacer
KNURLED ALUMINUM
LOCKING NUT
chrome-plated
sliding hood
&
buttcap
straight, NS,
Rolled welt, drawn
ferrules are
wrapped-over

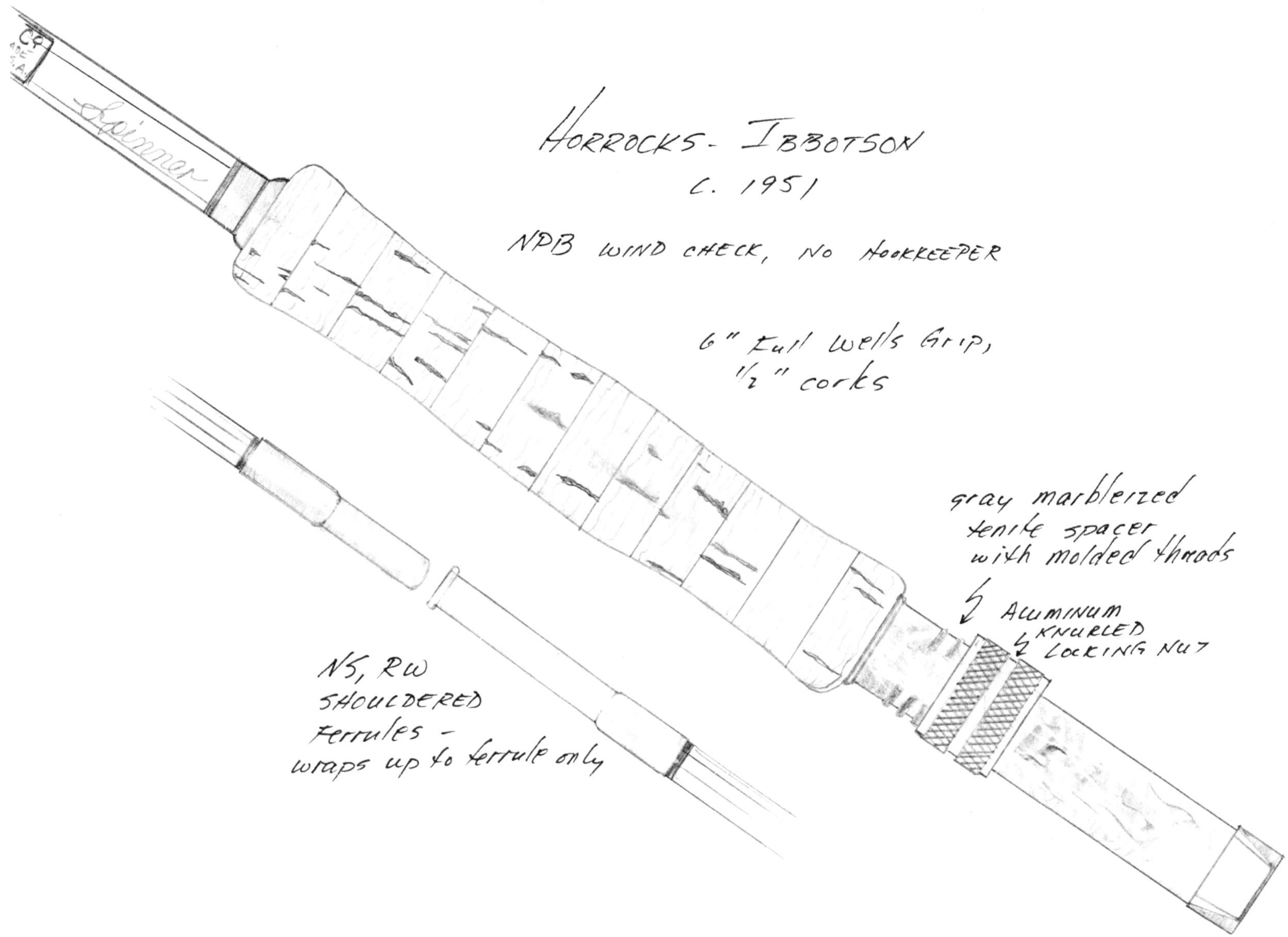
Spinner
HORROCKS-IBBOTSON
c. 1951
NPB WIND CHECK, NO HOOKKEEPER
6" Full Wells Grip,
1/2" corks
gray marbleized
tenite spacer
with molded threads
ALUMINUM
KNURLED
LOCKING NUT
NS, RW
SHOULDERED
Ferrules -
wraps up to ferrule only

Horrocks–Ibbotson Standard Reelseats

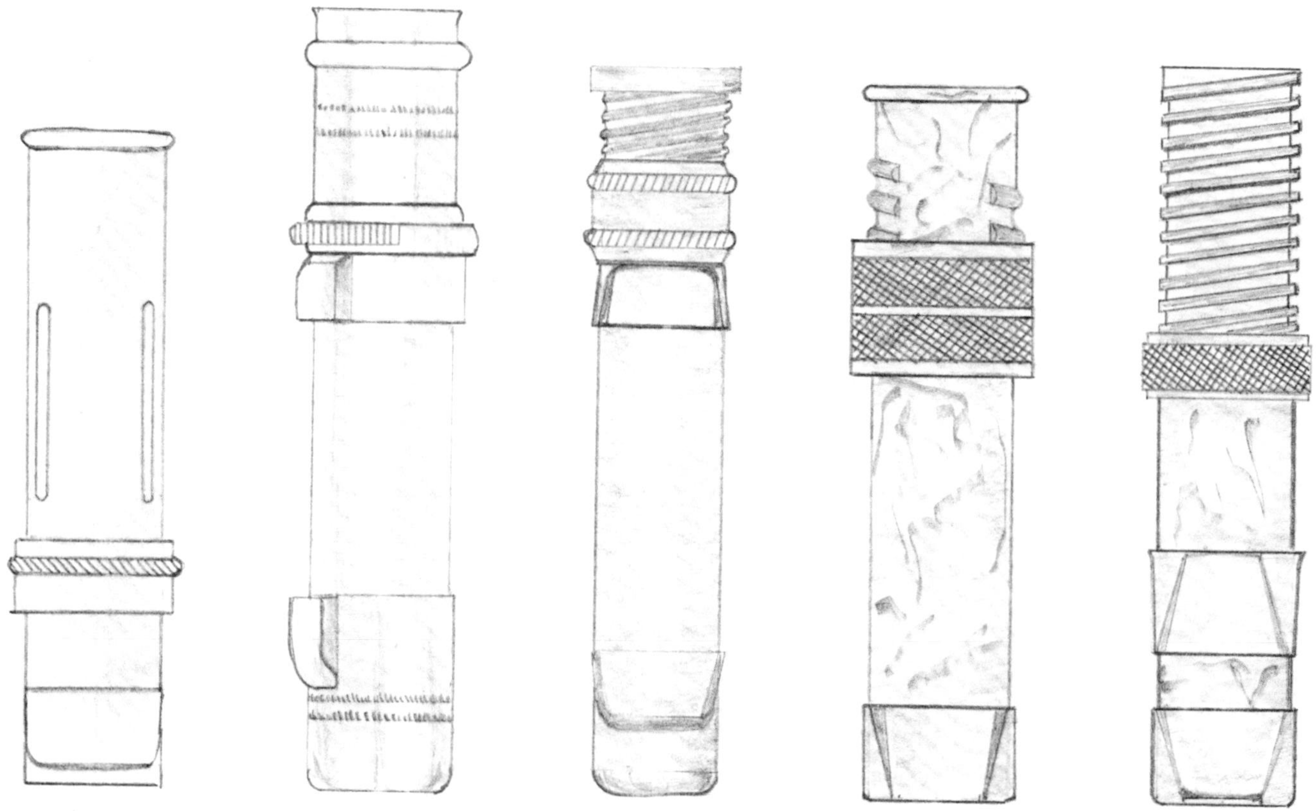

Montague Rod & Reel Company

If you are wondering which company had the longest history and made more rods in more grades than any other, the answer is Montague. Eugene P. Bartlett and Lester Leander Bartlett had been making fishing rods in Pelham, Massachusetts, since at least 1857, years before H.L. Leonard ever planed a piece of cane.

L. L. Bartlett left his brother's business in 1882, and in 1885 incorporated the Montague City Rod Company with two partners, Charles W. Hazeltin and Bernard W. Farren. The Montague City Rod Company began to buy out other tackle firms - first was Bartlett's brother Eugene in 1889, followed by the acquisition of The Thomas H. Chubb Rod Co in 1890. In about 1895 they bought the Malleson Rod and Reel Co of Brooklyn, N.Y. and thus had four rodmaking plants. In the mid 1920's to early 1930's, a consolidation was accomplished, and the plant at Montague City was enlarged several times to accommodate the machinery of the other three factories.

By about 1930, the name of the company was changed to the Montague Rod & Reel Company under the ownership of W. Bruce Pirnie who bought the stock owned by the original owners' descendants, and sold stock to the public. In 1934, the company was bought by Paul J. Johnson and Lee Moskowitz. When bamboo production was stopped in 1955, the bamboo stocks and machinery were purchased by Sewell N. Dunton and moved to Greenfield, Mass. Since most of the manufacturers of fishing rods were switching to fiberglass, Dunton was able to buy bamboo, ferrules and other hardware from many companies including H-I, Union Hardware, and South Bend. Dunton rods were marked with an oval decal on the rodshaft. Because they were often built of parts from many makers, they are difficult to identify if the decal has been removed. In 1974, Sewell Dunton sold his rodmaking operation to Thomas Dorsey and Thomas Maxwell who had joined to form Thomas & Thomas.

From c. 1915 to c. 1933, George I. Varney was the head rodmaker. He was responsible for Montague's highest quality rods. In an unprecedented agreement, Varney was allowed to produce very high quality rods marked "George I. Varney / Maker / Montague City, Mass." These rods used Varney's patented serrated ferrules and the highest grade NS hardware from the Chubb factory. Varney rods have very complex and time-consuming wrap patterns. See the drawing of the c.1920 Varney rod for details and typical wraps.

All of the phases of ownership and the large number of operations that made up the Montague Rod & Reel Company (MR&RC) help to explain the endless variety of rods made by the company throughout its 100+ year history. As each of the companies was integrated into MR&RC, different levels of quality were added to use the supplies of hardware and cane that each new acquisitions brought along.

Key among these was the Thomas H. Chubb Company of Post Mills, Vermont. This venerable firm had been in existence for many years, and had huge stores of hardware of its own manufacture. Chubb hardware is usually ornately knurled, and embellished in many ways. One example is the *sculptured* reelfoot guides on the NS reelseats. (See the drawing of the Chubb Spliced Rod in the chapter on Antiques). These are certain signs of Chubb pedigree, but are usually seen on rods prior to the Montague connection. Chubb invented the procedure for drawing nickel silver into a seamless, waterproof tube, which was patented for reelseats in 1880. Soon, this method was applied to the production of ferrules. Ferrules by Chubb have a few unique features. They have a larger welt than most ferrules, with a very distinctive shape. See the drawing for details.

Also, most Chubb ferrules have two rows of knurled *slashes* for ornamentation near the serrated end. From 1890 until 1932, Chubb rods continued to be made in Vermont and were marked and marketed as Chubb rods. These were Montague's pride - much as the Cross rods marketed by South Bend were their higher grade rods. They are referred to as Chubb/ Montague Rods. See the drawing of the Chubb c. 1920 Silkien Rod for a fine example of a higher quality Chubb. After 1920, some of the Chubb/Montague rods were of lower grades, using the older Chubb hardware and sheet cork grips. Many of these used reelseats that were produced 20 years earlier, making them appear to be older than they actually are.

Also, many of these were sold as trade rods in this era, and may have had unfamiliar markings. Usually these were of middle to lower quality, and used the bottle-shaped, shouldered ferrule. See the drawing for details. By 1930, the Chubb rods were of just about every quality level. For a great example of a mid-grade Chubb trade rod, see the drawing of the c. 1930 Chubb marked with the *Diamond Edge* logo. Diamond Edge was a trade name used by the Norvell-Shapleigh Hardware Company of St. Louis.

Rods sold by Montague and carrying their name were of many quality levels. Those that were listed in their catalogs of the 1920's and 1930's were comparable to almost every maker of that era. The rods made as trade rods were of middle to very low quality to meet the demands of the trade buyers. For a good example of a middle quality Montague of the 1920's, see the drawing. Note the straight, rolled welt ferrules with no serrations. The ferrules and other hardware are made of yellow brass that was oxidized. Montague called this *Bronzing*. Note, too, that the windings come up to the bottom of the ferrule, and do not *wrap-over* the bottom of the ferrule. This rod came put up on a form case that fit into a canvas bag. Higher quality rods would have NS reelseats, hand welted, NS ferrules that were serrated and wound-over, and better packaging.

The core line of Montague rods was consistent from the mid 1930's through the 1950's, with additions and subtractions made to meet the demands of the buyers. Montague catalogs were proud to state that *Montague Rods are sold in virtually every city and town in America*. They also had a rod to suit every pocketbook, with prices ranging from $2 to $50 in the 1940's. The basic line was introduced in the mid 1930's. Those model names were consistent throughout the rest of the bamboo era. These core-line models changed little throughout the years, and were the only rods that had fairly consistent wrap colors from about 1940 on.

Many of the Montague rods were wrapped by women who worked in their homes, and they wrapped many rods to suit themselves - when they got tired of wrapping red tipped in black, they might change to black tipped in red, or to any other combination that they fancied. The company allowed this and accommodated the ladies by advertising the rods only as *wrapped in two colors of silk.*

For the Montague wrap charts in this book, note that when colors are specified, it is because either these were specified in advertising and catalogs or because a significant number of examples have been seen wrapped as shown. Unless a rod's wrap colors are specified in the charts, you really cannot go too far wrong with any pleasing combination of colors. You can't do any worse than one of Montague's combinations: purple tipped red.

Montague hardware is easily identified. The earliest style of winding check was a *Bell-shaped* check usually made of yellow brass. Sometimes it was bronzed, sometimes nickel plated. The next style, and the longest lived, was a *Stepped Cone* usually with a couple of engraved lines for ornamentation. The latest was a wide, plain, flattened cone that was used on all models. This check was almost always nickel-plated brass. Most Montague rods carried a hookkeeper of the bridle and ring type. Usually it was mounted on the #6 flat and was wrapped in the closed style.

Ferrules used on Montague rods were usually one of three styles, but were made in many quality levels. The best ferrules were straight, hand welted, serrated, waterproof and designed to have the thread wrapped-over the serrations and the base of the ferrule. These often had two incised bands just above the wraps for ornamentation. The best of this style were made of NS, sometimes oxidized, bronzed or chrome plated. Some models had a *Satin-Chrome Finish* which is a chrome-plated NS ferrule that is treated to have a *flat* finish, rather than a highly polished one. These are often seen on pre-war Redwing models. The second style of ferrule, used on middle-grade rods is a straight ferrule with a rolled welt, *no serrations*, and no wrap over the base of the ferrule. These are from 1930 or earlier. Finishes ranged from plain NS to yellow brass that was bronzed, chrome-plated or nickel-plated.

The last style of ferrule is the *Bottle-Shaped*, or *Shouldered* ferrule. Montague was the leader in use of this ferrule style both in number of rods that carried it and the length of time it was used. The Montague style shouldered ferrule is easy to differentiate from those of other makers. See the drawing for details. Horrocks-Ibbotson used a shouldered ferrule of similar but much simpler and cheaper design. If you see the two side-by-side, you will have no trouble in determining which is the Montague. Montague shouldered ferrules were almost always yellow brass plated with Chrome or Nickel, with chrome plating being the *better* ferrule. Near the base of the ferrule was a slight *waist*. This indicates its Montague origin.

There was a seldom used Montague ferrule that was a shouldered, chrome-plated ferrule with no waist, no ornamentation and *no welt* at all. This was used on the very, very cheapest of all trade rods and only on the two cheapest cataloged Montague rods. It looks like an H-I ferrule, except that H-I shouldered ferrules were not chrome-plated. There is a third rarely seen Montague ferrule that is actually a vestige of the old Chubb Rod Company. Thomas Chubb was fascinated with every metal part of the rod, and patented more hardware parts than any other rodmaker. One of his most interesting ideas was to make a ferrule with a *"Hexagonal Base*. Remember that in ferrule design, it is best not to have to remove any more of the cane at the ferrule station than is absolutely necessary. This being true, Chubb's Hex-Base ferrule should have been the ultimate answer. In theory it was a sound idea, but in practice the lack of serrations was more likely to damage the cane at the edge of the ferrule than the hex shape was to protect it.

Hex-Base ferrules were still in use as late as the 1940's on the two highest grades of Montague's Powr Built Series. These were double-built rods as made famous by Cross Rod Company. Perhaps these ferrules were used on these rods not because the ferrules were considered to be *highest-grade* but because the inherently stronger double-built rods were much less likely to suffer a stress fracture. For details on this ferrule, see the drawing of the spliced Thomas Chubb Silkien Rod in the chapter on antiques.

Cane work on Montague rods can vary as widely as the quality of hardware. Higher grade rods - Manitou, Redwing, Trail and Fishkill – are as good as any higher-grade production rods. Grades from Rapidan to Sunbeam are strictly average.

Reelseats throughout the years were as varied as the different models. The lowest grade rods and those made before 1906 featured sliding band type reelseats. These were usually very plain and made of nickel-plated brass. Sometimes the yellow brass was treated and was called *Blackened Brass* or sometimes *Bronzed Brass*. When this treatment was used on the reelseat, the ferrules were usually treated in the same manner. If you find a Montague rod with a NS sliding band seat, it is likely to be pre-1906 and of high grade. Such a rod would probably have a swelled butt. Most of the high grade sliding band seats were of Chubb origin and resemble the seat shown on the c. 1920 Diamond Edge rod in the drawing.

Montague continued to modify its sliding band reelseat through the years, but it never changed too drastically. The proportions became more refined over the years and the dimensions became smaller until the version shown in the drawing was the most common design for the lower grade rods. If the reelseat was stamped MONTAGUE as shown in the drawing, the rod was sold under the Montague name. Reelseats of this style without the Montague marking were sold as trade rods.

The Montague Style Locking Band reelseat is the result of four patents by Leander Bartlett. The earliest patent was in 1904, the remaining three were granted in 1906, so that year is usually seen if any patent date is stamped on the locking band. This reelseat was used from 1906 to the end of WW II. It is very easily recognized by the arrow and slot on the two-piece locking band. To release the locking band, align the arrow and the slot in the knurled band, then slide the locking band to its desired position and turn the knurled band so that the arrow is out of alignment with the slot. This will lock the band in place. Though the style of the barrel or spacer used for this reelseat changed through the years, the locking band itself looked essentially the same. After c. 1930, this reelseat was more popular on larger rods such as boat, pier or trolling rods as it was a very secure but heavy reelseat.

By 1939, the reelseats had entered the *modern* era and began to feature screw, down-locking mechanisms and plastic composite spacers. Lower grade rods still used sliding band seats as in the past, but *low-medium grade* rods were equipped with a new style called the Titelock reelseat. This style featured a wide, knurled NPB locking band and an independent sliding band. The spacer material varied. Lowest grade rods used an NPB spacer, next highest was an aluminum spacer, a mid-grade Titelock would feature plastic threads and spacer, and the highest grade Titelock had aluminum threads and a cedar spacer.

After WW II the Titelock reelseat was changed slightly. The locking band was virtually the same, but the sliding band was replaced with a band that had a reelfoot hood. This band was connected to the knurled locking band. All of the metal was chrome plated brass, and the spacer was always plastic – usually blue, red, brown or green to compliment the wrap colors. A third version of the Titelock was used on higher grade post-war rods. It had longer threads and a short knurled locking band.

The better Montague rods from c. 1933 until the end of bamboo production usually featured the Scrulock Reelseat. This seat had a narrow knurled locking band made of aluminum and an independent sliding band made of nickel-plated brass. Earlier Scrulock seats were usually fitted with nickel-plated brass spacers and hardware. A later and somewhat better version had threads and spacer of plastic and chrome-plated brass hardware. The spacer was very often white or black plastic. The best Montague reelseat was a fancy version of the Scrulock seat. It had NS hardware (sometimes blackened or bronzed) and a nicely machined sliding band. The spacer material varied with the era and quality level of the rod. Cedar, plastic, hard rubber and Tenite were the most common. This reelseat is high grade in all ways.

Montague Decals

The earliest decal was round. This decal was used c. 1910 to 1932. It was about ½" in diameter and featured a large, black capitol "M" on a white background. Overlaid on the "M" is an oval plaque showing the image of an eagles's talon gripping a fish. Often, the white background appears transparent, especially if the varnish is in good condition.

The second decal was short lived, and may have been used only c. 1934 to c. 1939. It was a fat oval, about 7/8" across its widest dimension. This decal features a fighting rooster, a *Game Cock*. Apparently this was to rival another fighting emblem - Heddon's *Rod with the Fighting Hear*t. The Game Cock decal featured the Montague name across the top and the model name across the bottom. The colors varied, but usually the Game Cock image and lettering are gold foil, and the background is usually deep red.

From about 1939 to about 1949, the third style decal was used. This is the most commonly seen decal, and is the only way that the model name can be determined without the original bag or tube. (Although wrap colors are a very good indicator). This decal is a slimmer oval than the Game-Cock decal, though it is about the same width. The image in the center is an eagle's talon gripping a fish overlaid on an oval shaped plaque. Again featured is the Montague name above the image and the model name below the image. This is a colorful decal. The large oval is gold foil. Lettering is black. The dark center oval is red, the fish is white with blue shading and the eagle's talon is bright yellow.

After 1950, the fourth decal was used. It is identical to the third design except that the decal is more generic. Rather than having a decal for each model, the decal simply stated Montague Rod & Reel Company. The model name appeared only on the bag tag or tube label, and wrap colors were much more dependable for determining the model. Some have speculated that with so many people involved in making so many rods, over so many years, Montague was consistent only in its inconsistency.

MONTAGUE

ROD AND REEL COMPANY

MONTAGUE CITY MASSACHUSETTS

MONTAGUE FLASH

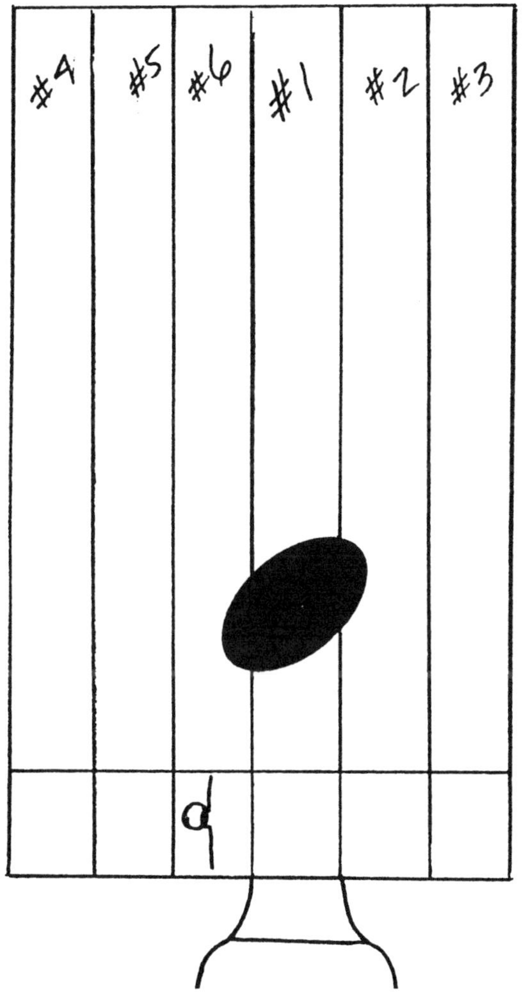

MONTAGUE GAME COCK RED WING

MONTAGUE ROD & REEL COMPANY

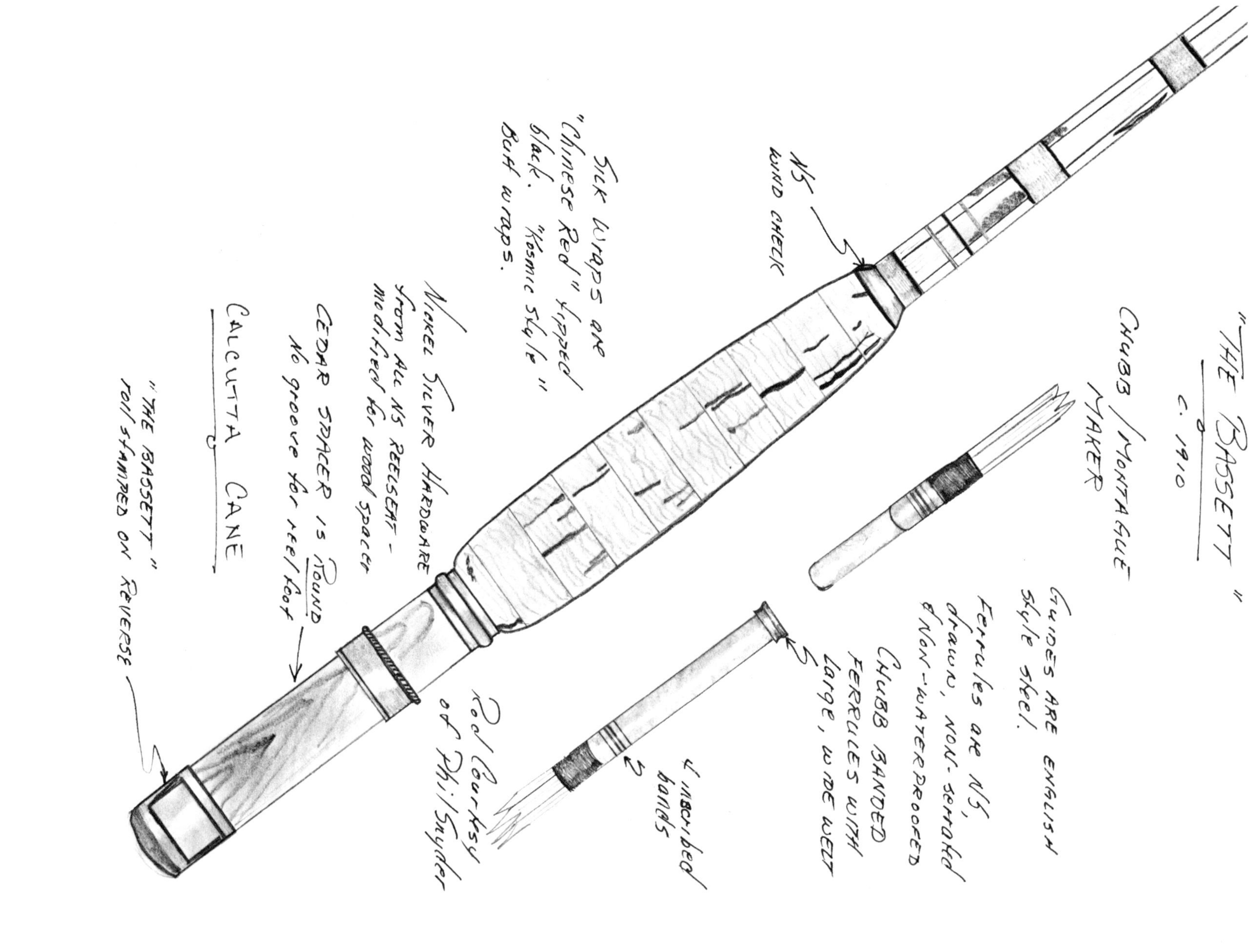
"THE BASSETT"
c. 1910
CHUBB / MONTAGUE
MAKER
GUIDES ARE ENGLISH
style steel.
Ferrules are NS,
drawn, NON-serrated
& NON-WATERPROOFED
CHUBB BANDED
FERRULES WITH
large, WIDE WELT
4 inscribed
bands
Rod Courtesy
of Phil Snyder
NS
WIND CHECK
SILK WRAPS are
"Chinese Red" tipped
black. "Kosmic Style"
Butt wraps.
NICKEL SILVER HARDWARE
from ALL NS REELSEAT -
modified for wood spacer
CEDAR SPACER IS ROUND
No groove for reel foot
CALCUTTA CANE
"THE BASSETT"
foil stamped on REVERSE

Montague City Rod Co. Hardware Specs 1920's

Model	$	Reelseat	Grip	Stripper	Snake	TT	Ferrules
Manitou	35.00	NS, Sprial-Lock, BHR	5½" Full Wells	A	HS	A	NS, HW, S, WP
Red Wing	25.00	NS, SB, S, Hood	5½" Rev. HW	A	HS	A	NS, HW, S, WP
Trail	20.00	NS, SB, S, Hood	5½" Rev. HW	A	HS	A	NS, HW, S, WP
Flipline	15.00	NS, SB, S, Hood	5½" Peckerhead	HS	HS	HS	NS, RW
Fishkill	12.00	NPB, SB, S, Hood	5½" Half Wells	HS	HS	HS	NPB, RW
Splitswitch	8.00	NPB, C&R	5½" Cigar	HS	HS	HS	NPB, RW
Fishkill Combo	13.50	NPB, C&R	Reversible Cigar	HS	HS	HS	NPB, RW

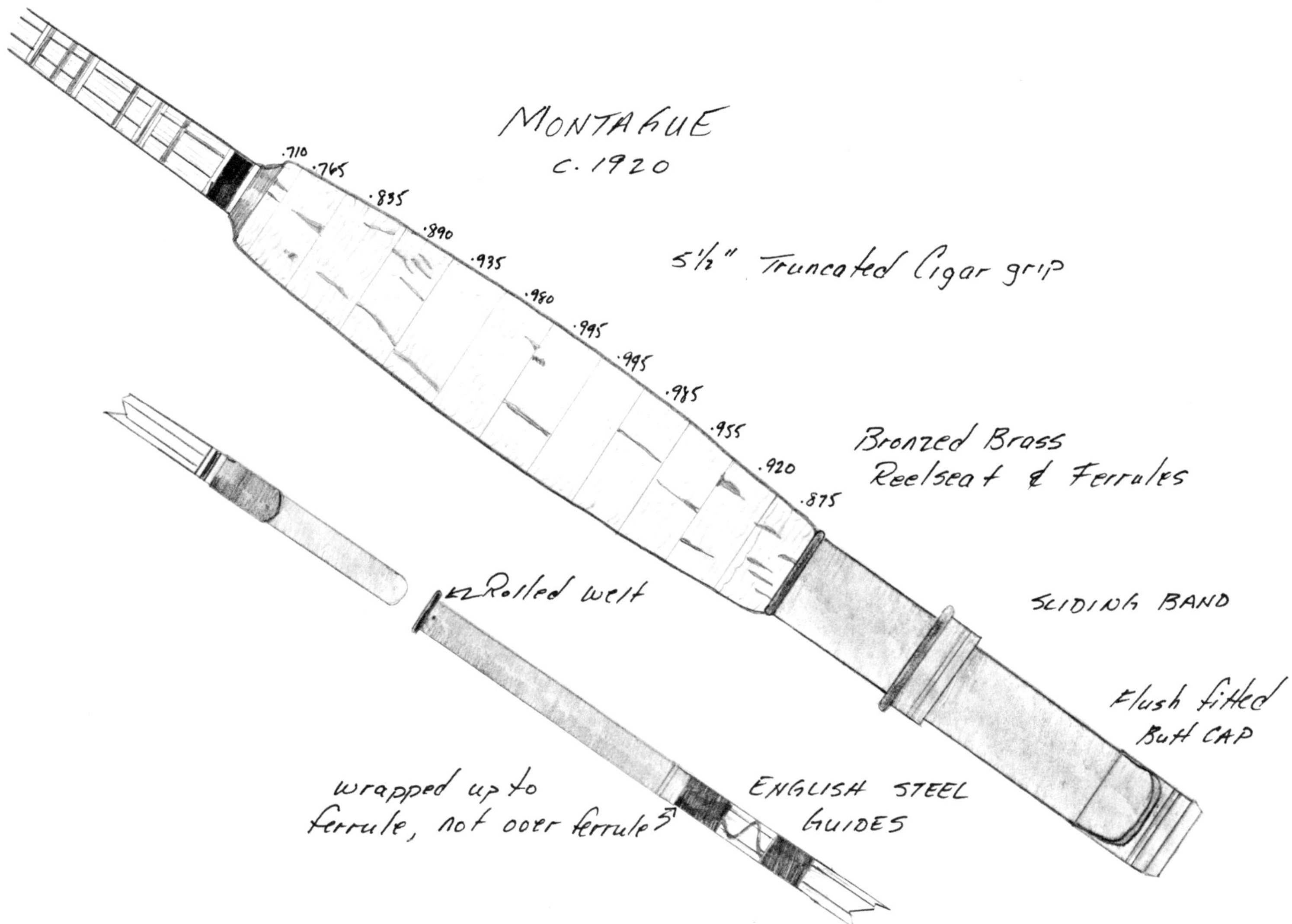
MONTAGUE
c. 1920
.710
.765
.835
.890
.935
.980
.995
.995
.985
.955
.920
.875
5½" Truncated Cigar grip
Bronzed Brass
Reelseat & Ferrules
Rolled welt
SLIDING BAND
Flush fitted
Butt CAP
wrapped up to
ferrule, not over ferrule
ENGLISH STEEL
GUIDES

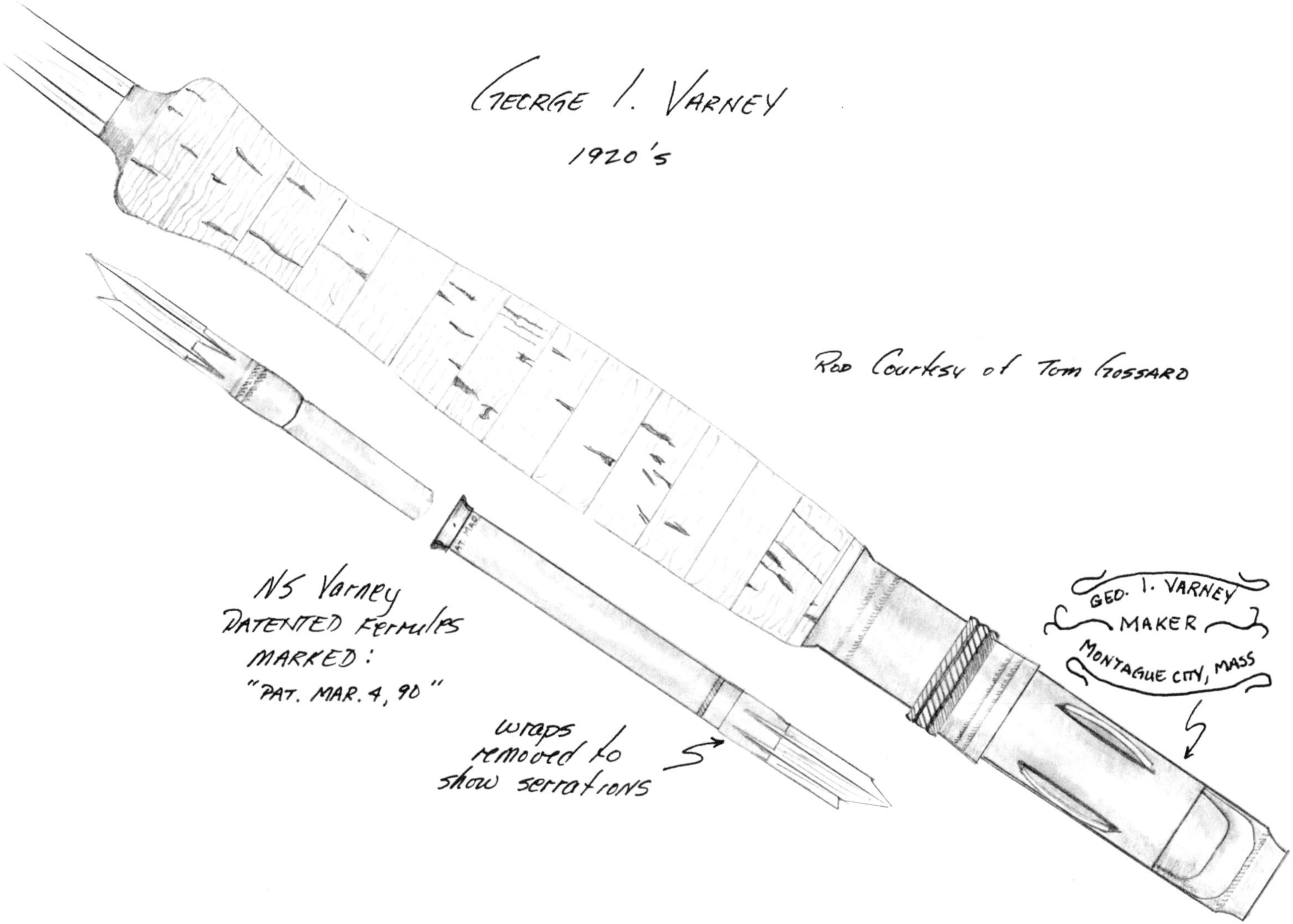

GEORGE I. VARNEY
1920's
Rod Courtesy of Tom Gossard
NS Varney
PATENTED Ferrules
MARKED:
"PAT. MAR. 4, 90"
wraps
removed to
show serrations
GEO. I. VARNEY
MAKER
MONTAGUE CITY, MASS

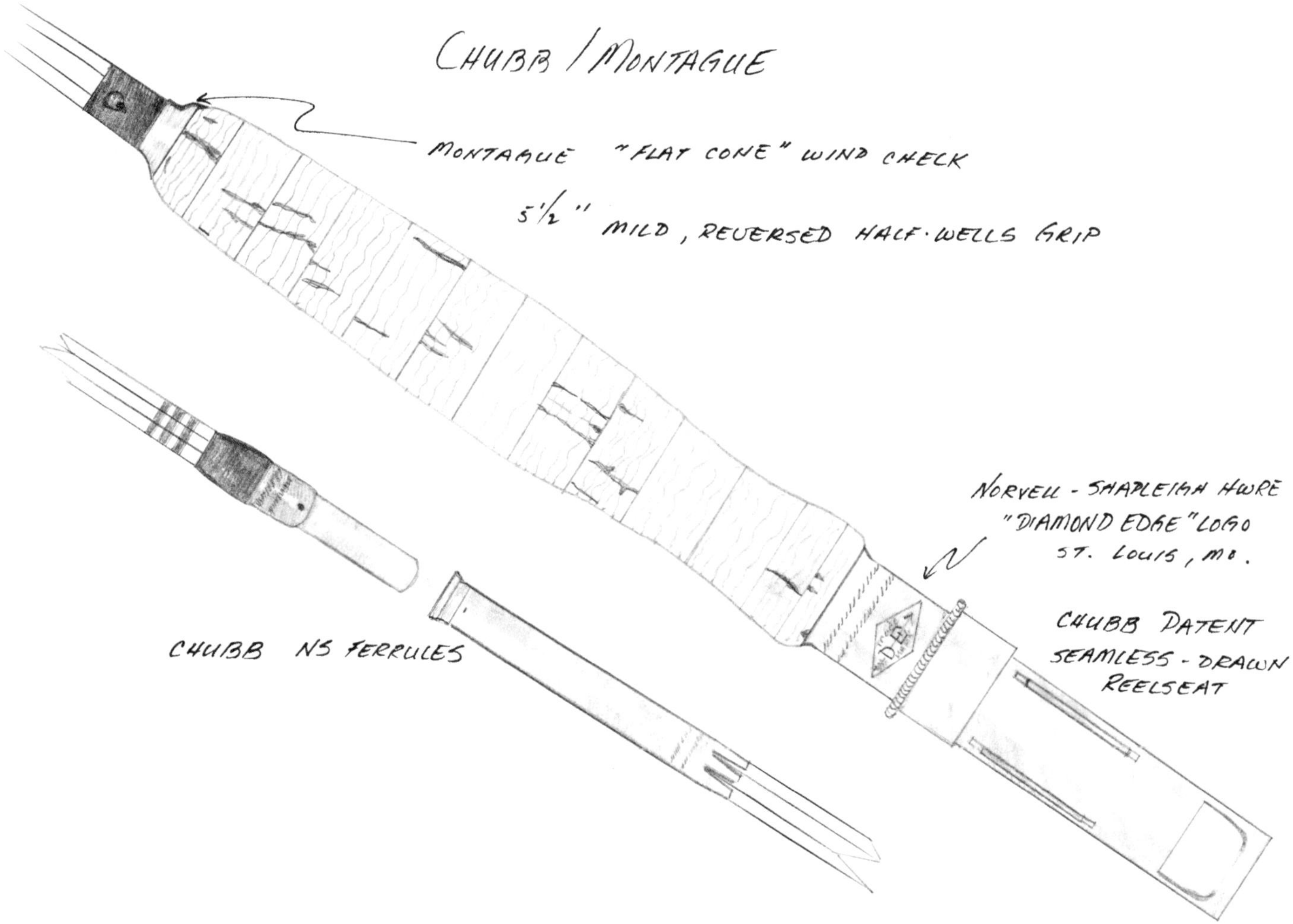
CHUBB / MONTAGUE
MONTAGUE "FLAT CONE" WIND CHECK
5½" MILD, REVERSED HALF-WELLS GRIP
NORVELL - SHAPLEIGH HWRE "DIAMOND EDGE" LOGO ST. LOUIS, MO.
CHUBB PATENT SEAMLESS - DRAWN REELSEAT
CHUBB NS FERRULES

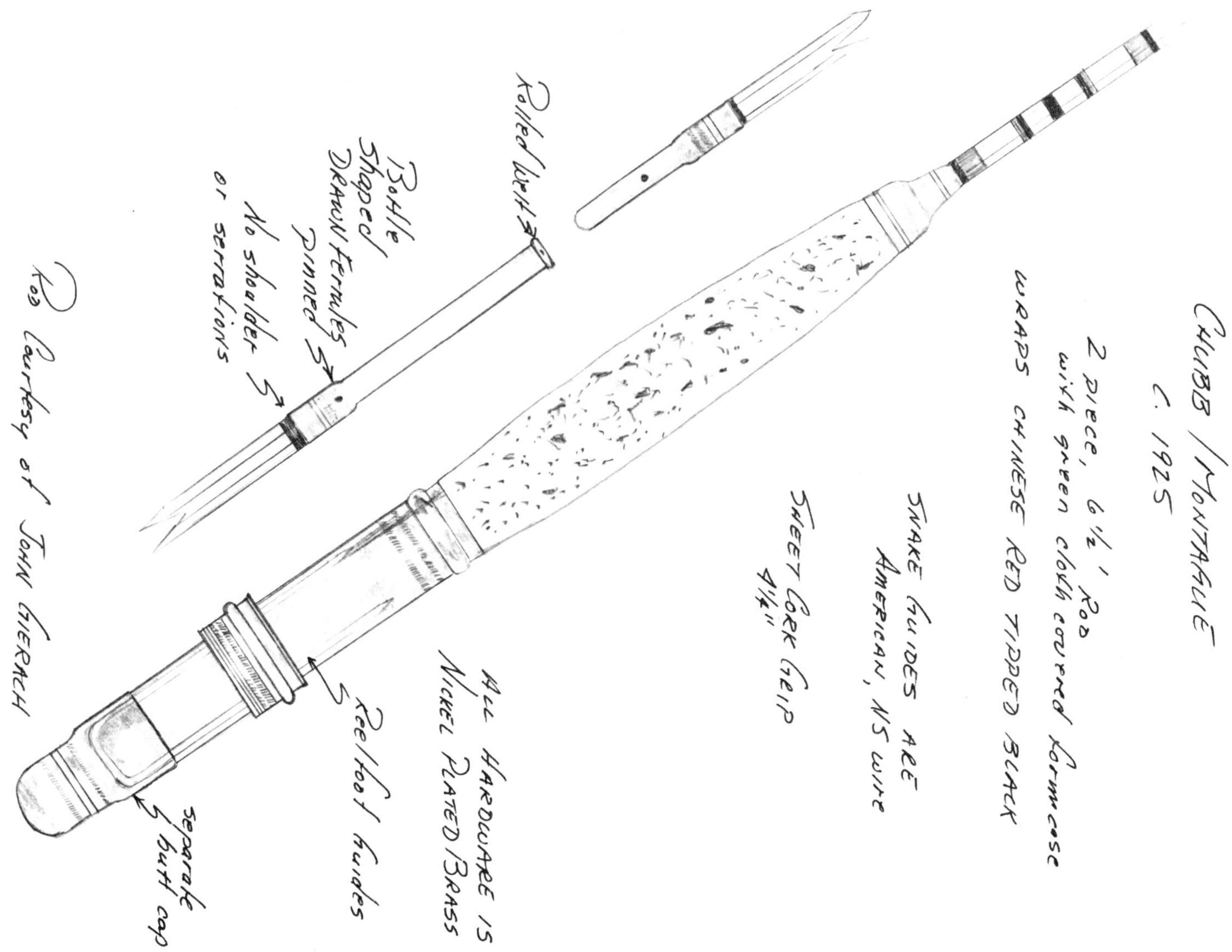
CHUBB / MONTAGUE
c. 1925
2 piece, 6 1/2' Rod
with green cloth covered form case
WRAPS CHINESE RED TIPPED BLACK
SNAKE GUIDES ARE
AMERICAN, NS wire
SWEET CORK GRIP
4 1/4"
ALL HARDWARE IS
NICKEL PLATED BRASS
Reel foot guides
separate
butt cap
Rolled welt
Bottle
Shaped
DRAWN ferrules
pinned
No shoulder
or serrations
Rod Courtesy of JOHN GIERACH

Montague Rod & Reel Co. Wrap Specs 1920's

Manitou
Pongee #27
Purple #16

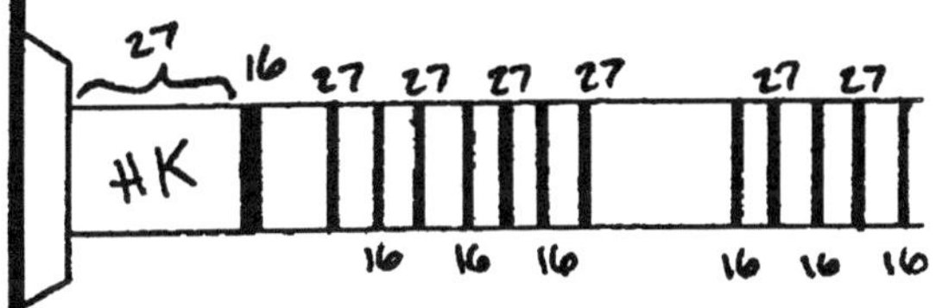

Signature

Red Wing
Red #10
Yellow #3

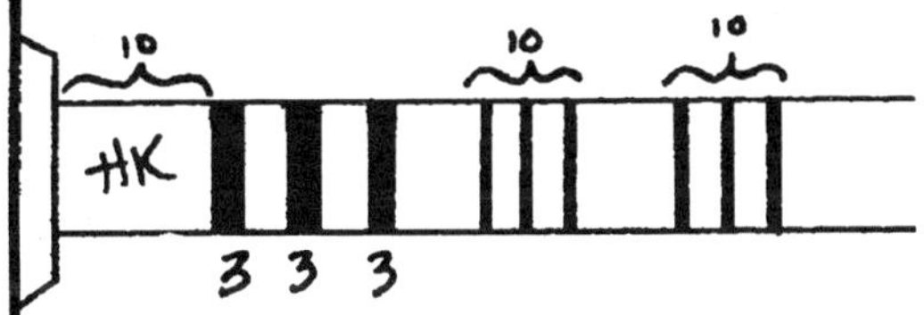

Signature

Trail
Yellow #3
Red #10
Yellow #3 Intermediates

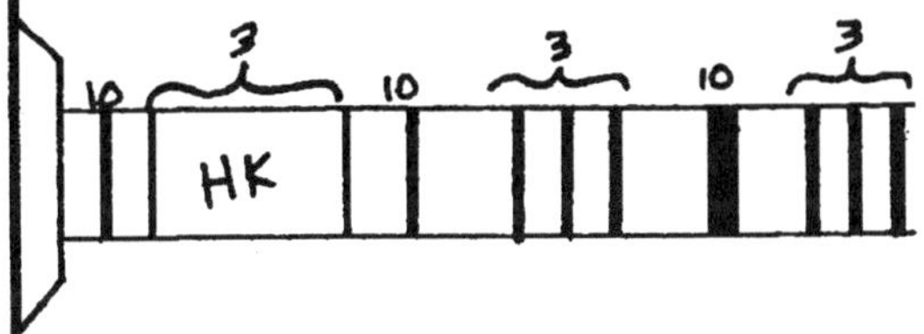

Signature

Flipline
Garnet #12
Lemon #4 for Tipping
Intermediates Garnet #12

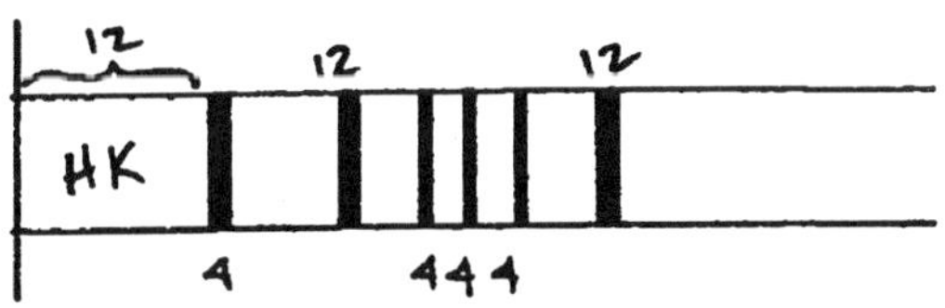

Signature

Fishkill
Jasper #X
Intermediates Jasper #X

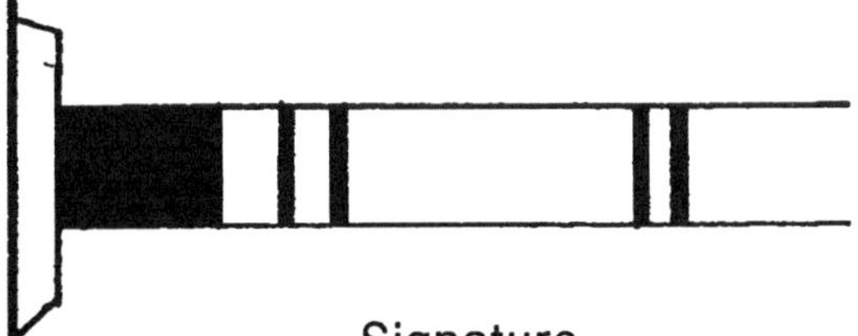

Signature

Splitswitch
Jasper #X
No Tipping
Intermediates Jasper #X

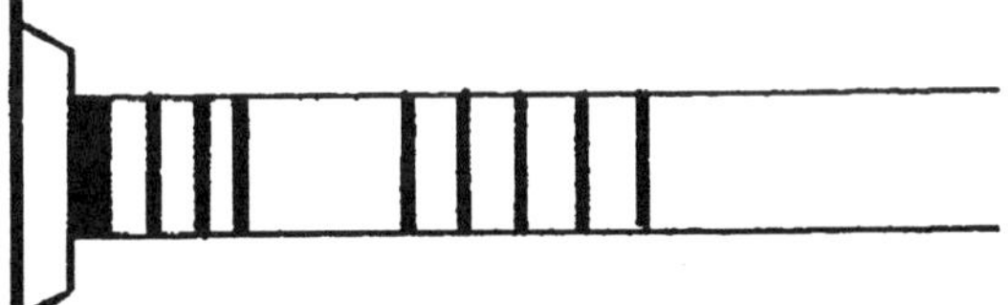

Signature

Montague City Rod Co. — Hardware & Wrap Specs — 1932

Model	$	Cane	Reelseat	Grip	Stripper	Snake	TT	Ferrules	Wraps	Tipping
Clipper	3.00	BK	NPB, C&R	Cigar	Snake	SS	SS	NPB, RW, SH	Red #10	None
Sunbeam	4.00	DBR	NPB, C&R	Cigar	Snake	SS	SS	NPB, RW, SH	Green #23	Red #10
Flash	5.00	None	NPB, C&R	Cigar	SS	BS	HS	NPB, RW, SH	Red #10	Green #23
Rapidan	6.00	Maroon	NPB, C&R	Half Wells	Ag	SS	HS	NPB, RW, SH	Red #10	Green #23
Splitswitch	8.00	None	NS, C&R	Cigar	Ag	SS	Ag/HS	NS, RW, SH	2 colors	
Fishkill	12.00	None	NS, C&R	Half Wells	Ag	SS	Ag	NS, RW, SH	2 colors	
Flipline	15.00	None	NS, C&R	Peckerhead	A	SS	A	NS, RW	Gold #6	Black #1
Trail	20.00	None	NS, C&R	Full Wells	A	SS	A	NS, HW, S	Yellow #3	Red #11
Comet	10.00	None	BNS, C&R	Cigar	Ag	NS	Ag	BNS, RW	Purple #16	Gold #6
Red Wing	25.00	None	NS, SB	Rev. HW	A	SS	A	NS, HW, S, P	Red #10	Yellow #3
Manitou	35.00	None	B, HR, SDL	Full Wells	A	SS	A	NS, HW, S, WP	Pongee #27	Purple #16

Montague Rod & Reel Co. Hardware Specs 1940's

Model	$	Cane	Reelseat	Grip	HK	Stripper	Snake	TT	Ferrules
Powr-Built 50	50.00	F	SDL, Tenite	Half Wells	Y	PB	PB	PB	NS, HW, S, Hex base
Powr-Built 35	35.00	F	SDL, Tenite	Half Wells	Y	PC	PC	PC	CPNS, HW, S, Hex base
Powr-Built 25	25.00	F	SDL, AL, Rubber	Half Wells	Y	PB	PB	PB	BNS, RW, S
Powr-Built 20	20.00	F	SDL, NS	Half Wells	Y	SS	SS	PC	CPNS, RW, S
Manitou Tour.	40.00	BN	NS, SB, SH	Mild Wells	Y	PC	TS	TS	NS, HW, S, WP, w/o
Manitou	35.00	F	SDL, BNS, HR	Mild Wells	Y	A	TS	A/PL	BNS, HW, S, WP, w/o
Red Wing	25.00	F	SDL, SCPNS, C	Mild Wells	Y	PC	PC	PC	SCP, HW, S, WP, w/o
Trail	20.00	F	SDL, CP, Comp.	Mild Wells	Y	A	TS	A/TS	NS, HW, S, WP, w/o
Flipline	17.50	F	SDL, CP, Cedar	Mod. Cigar	Y	A	TS	A/TS	NS, HW, S, WP, w/o
Fishkill	15.00	F	SDL, BNS, Cedar	Mild HW	Y	Ag	TS	Ag	BNS, RW, S, WP, w/o
Comet	12.00	F	SDL, CPB, Scrulock	6" Cigar	Y	CP	CP	CP	CPNS, RW, S, WP, w/o
Splitswitch	10.00	F	SDL, CPB, Scrulock	Cigar	Y	Ag	TS	Ag/TS	NS, RW, S, WP, w/o
Broad Brook	9.75	F	SDL, BNPB, Scrulock	7" Str.	Y	Ag	TS	Ag/B	BPN, RW, SH
Blue Streak	12.00	F	SDL, CPB, Scrulock	7" Str.	Y	CP	CP	PC	CPNS, HW, SH
Rapidan	8.50	F	SDL, BNPB, Scrulock	Cigar	Y	Ag	TS	Ag/TS	BPN, RW, SH
Flash	6.25	F	MLB, MPB	6" Wells	Y	TS	TS	TS	NPB, RW, SH
Leaping Brook	5.25	F	MLB, NPB	6" Wells	Y	HS	TS	HS	NPB, RW, SH
Beaver Run	7.00	F	MLB, NPB	6" Wells	Y	Ag	TS	Ag/HS	NPB, RW, SH
Madison	8.25	F	CPB Scrulock	6" Wells	Y	SS	SS	PC	CPB, RW, SH
Commanche	6.75	F	CPB Scrulock	6" Wells	Y	HS	TS	HS	CPB, RW, SH
Lakeside	6.75	F	NP Titelock	6½" HW	Y	A	TS	Ag/HS	NPB, RW, SH
Apache	6.25	F	NP Titelock	6½" HW	Y	A	BS	Ag/BS	NPB, RW, SH

Continued on Next Page

Montague Rod & Reel Co. — Hardware Specs — 1940's

Model	$	Cane	Reelseat	Grip	HK	Stripper	Snake	TT	Ferrules
Ardacian	5.25	F	MLB, NPB	5½" HW	Y	A	TS	Ag/HS	NPB, RW, SH
AuSable	5.00	F	MLB, NPB	5½" HW	Y	SS	TS	HS	NPB, RW, SH
Rainy River	4.25	F	NPB, C&R	5½" HW	Y	Snake	TS	HS	NPB, RW, SH
Mountain Lake	7.00	BN	CPB, Titelock	6½" HW	Y	Ag	TS	Ag/HS	CPB, RW, SH
Pelham Brook	6.00	BN	CPB, Titelock	6½" HW	Y	TS	TS	HS	CPB, RW, SH
Bear Lake	5.75	BN	NPB, Titelock	6½" HW	Y	Ag	TS	Ag/HS	NPB, RW, SH
Sunbeam	5.00	BN	NPB, Titelock	6½" HW	Y	Snake	TS	HS	NPB, RW, SH
Clipper	4.50	BN	NPB, Titelock	6½" HW	Y	Snake	TS	HS	NPB, RW, SH
Cold Spring	5.25	BN	NPB, Titelock	6½" HW	Y	Ag	SS	Ag/HS	CP, RW, SH
Sonora	5.25	BN	NPB, Titelock	6½" HW	Y	Ag	TS	Ag/HS	NP, RW, SH
Deerfield	4.25	BN	NPB, C&R	6½" HW	Y	Ag	TS	Ag/HS	NP, RW, SH
Barton Brook	4.25	BN	MLB, NP	6½" HW	Y	Ag	TS	Ag/HS	NP, RW, SH
Eagle River	4.25	BN	NPB, C&R	7" Straight	Y	Snake	TS	HS	NP, RW, SH
Timberlake	3.75	BN	MLB, NP	6" Straight	Y	Snake	TS	HS	NP, RW, SH
Lake Pleasant	4.00	BN	NP, C&R	6" Cigar	Y	Snake	TS	HS	NP, RW, SH
Clear Lake	3.00	BN	NP, C&R	6" Cigar	Y	Snake	TS	HS	NP, RW, SH
Mt. Tom	3.50	BN	NP, C&R	6" Cigar	Y	Snake	TS	HS	NP, RW, SH
Highland	2.75	BN	NP, C&R	6" Cigar	Y	Snake	TS	HS	NP, RW, SH
11580/11579	3.25/2.50	BN	NP, C&R	6" Cigar	N	Snake	BS	HS	CP, NW, SH
500½/0500½	2.75/2.00	HT	NP, C&R	5½" Cigar	N	Snake	BS	HS	CP, NW, SH

Continued on Next Page

Montague Rod & Reel Co. Hardware Specs 1940's

Model	$	Cane	Reelseat	Grip	HK	Stripper	Snake	TT	Ferrules
Zephyr Brand Extra Light Rods									
Little Smoky	8.00	F	NPB Titelock	5½" Cigar	Y	Ag	TS	Ag/HS	NP, RW, SH
Sierra	7.00	F	NPB Titelock	5½" Cigar	Y	Ag	TS	Ag/HS	NP, RW, SH
Classon Brook	5.50	BN	NPB, MLB	5½" Cigar	Y	Ag	TS	Ag/HS	NP, RW, SH
Midge	4.50	BN	NPB, C&R	5½" Cigar	Y	Snake	TS	HS	NP, RW, SH
Celebrity Line of Fine Rods									
Winthrop	35.00	HT	Scrulock, BNPB	6½" MRHW	Y	PS	TS	PS	BNS, HW,S. WP, w/o, Hex
Bradford	25.00	HT	Scrulock, CPB	6½" MRHW	Y	CP	CP	PC	CPB, HW, S, WP, w/o Hex
Webster	20.00	HT	Scrulock, BNPB	6½" MRHW	Y	PS	BS	PBR	BNS, HW,S. WP, w/o, Hex
Palmer	15.00	HT	Scrulock, CPNS	6½" Cigar	Y	CP	CP	PC	CPNS, RW, S, WP, w/o
Churchill	10.00	HT	CPNS, C&R	6½" Cigar	Y	HS	TS	HS	CPNS, RW, S, WP, w/o
Two–Piece Fly Rods									
Manitou	35.00	F	Scrulock, BNPB	6½" Wells	Y	A	TS	A/PC	BNS, HW, S, WP, w/o
Fishkill	15.00	F	Scrulock, BNPB	6½" Wells	Y	A	TS	A/PC	BNS, HW, S, WP, w/o
Rapidan	8.50	F	NPB, C&R, Cedar	6½" Wells	Y	SS	SS	PC	NP, RW, WP, w/o
Pocono	7.00	F	Titelock, NPB	6½" Mild HW	Y	Ag	TS	Ag/HS	NP, RW, WP, w/o
Sunbeam	5.50	BN	NP, C&R	6½" Mild HW	Y	Snake	TS	HS	NP, RW, SH

Montague Rod & Reel Co. **Hardware & Wrap Specs** **1952**

Model	$	Reelseat	Grip	Stripper	Snake	TT	Ferrules	Wraps	Tipping
Amateur 2/1	7.65	NPR, C&R	Cigar	Snake	SS	SS	NPB, RW, SH	Red #10	None
Sunbeam 2/2	13.95	CP, ST, Blue	MHW	S	SS	SS	CPB, RW, SH	Blue #17	White #2
Rapidan 2/2	18.75	CP, ST, Black	MHW	PC	SS	SS	NS, RW	Brown #28	Gold #7
Fishkill 2/2	29.95	CP, ST, Green	MHW	PC	SS	SS	NS, HW, S	Green #24	Gold #7
Clear Lake	8.75	NPB, C&R	Cigar	Snake	SS	SS	NPB, RW, SH	Red #10	Green #24
Highland	9.85	NPB, C&R	Cigar	Snake	SS	SS	NPB, RW, SH	Green #24	None
Lake Pleasant	10.65	NPB, C&R	Cigar	S	SS	SS	NPB, RW, SH	Red #10	Purple #26
Mt. Tom	11.95	NPB, MLB	Straight	S	SS	SS	NPB, RW, SH	B&W Jasper	
Clipper	11.95	NPB, MLB	Straight	S	SS	PSS	NPB, RW, SH	B&W Jasper	
Sunbeam	13.95	CP, ST, Blue	MHW	PSS	SS	PSS	CPB, RW, SH	Blue #17	White #2
Rainy River	14.85	CP, ST, Red	MHW	PSS	SS	PSS	CPB, RW, SH	Gold #7	Red #10
Flash	16.50	CP, ST, Red	MHW	PSS	SS	PSS	NPB, RW, SH	Red #10	White #2
Rapidan	18.75	CP, ST, Black	MHW	PC	PC	PC	CPB, RW, SH	Brown #24	Gold #7
Blue Streak	23.60	CP, ST, Blue	MRHW	PC	PC	PC	NS, RW	Blue #17	Gold #7
Fishkill	29.95	CP, ST, Green	MHW	PC	PC	PC	NS, HW, S, WP	Green #24	Gold #7
Red Wing	41.40	CP, SDL, Red	Rev. HW	PC	PC	PC	NS, HW, S, WP	Red #10	Gold #7
Manitou	53.25	CP, SDL, Wood	Full Wells	A	PC	PC	NS, HW, S, WP	Pongee #27	Purple #16

Montague Standard Cork Grips

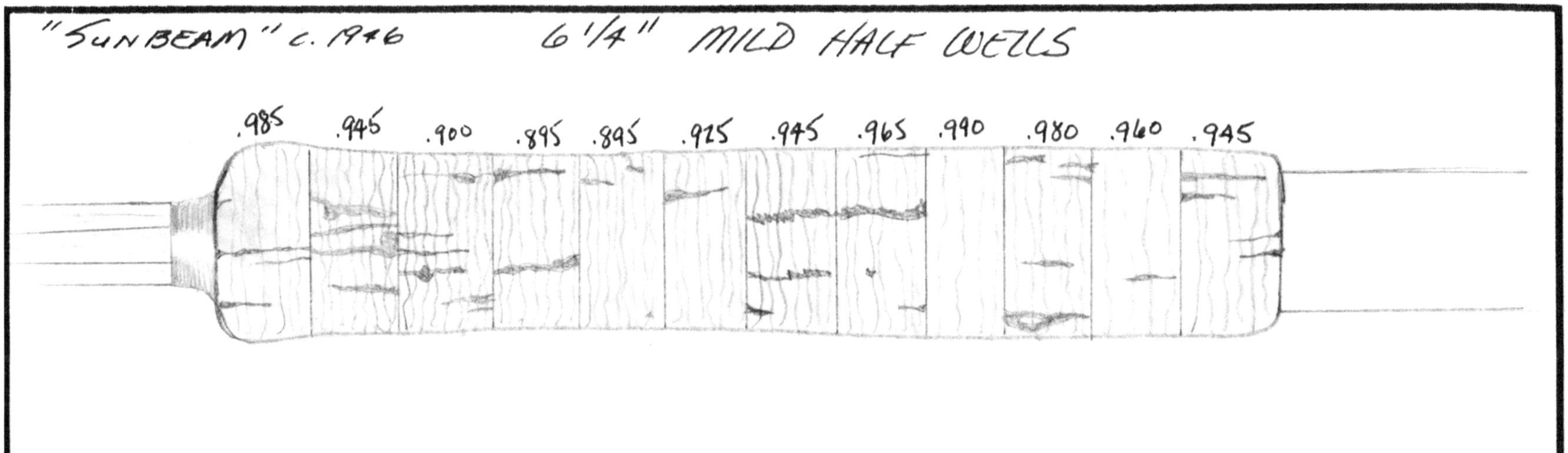

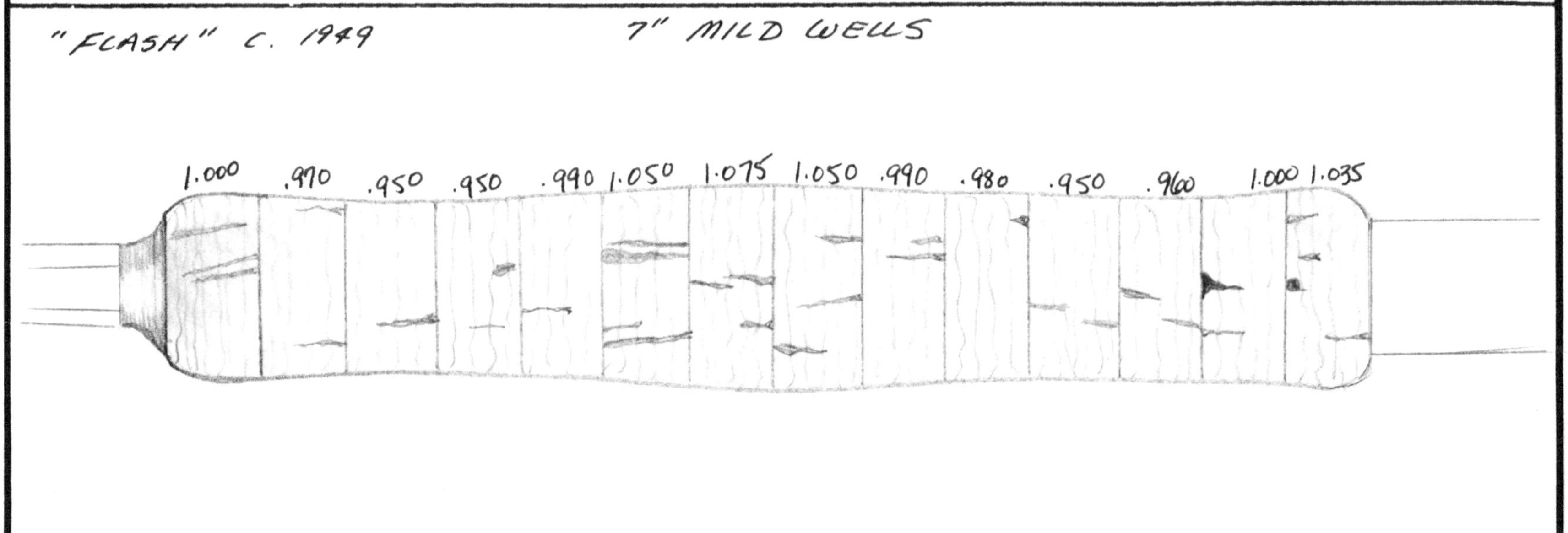

Montague Cork Grip and Reel Seat Variations

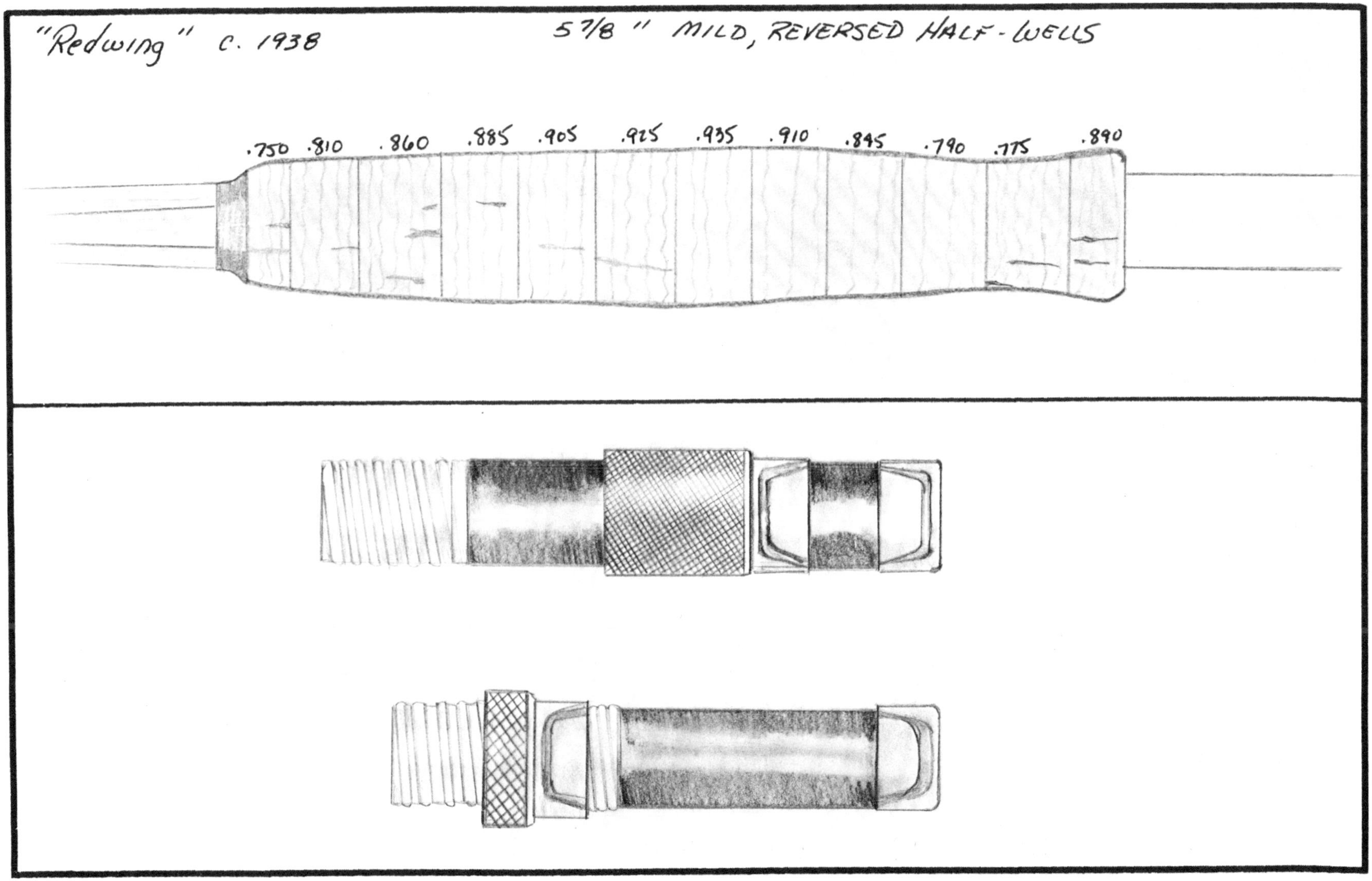

Montague Reel Seat Variations

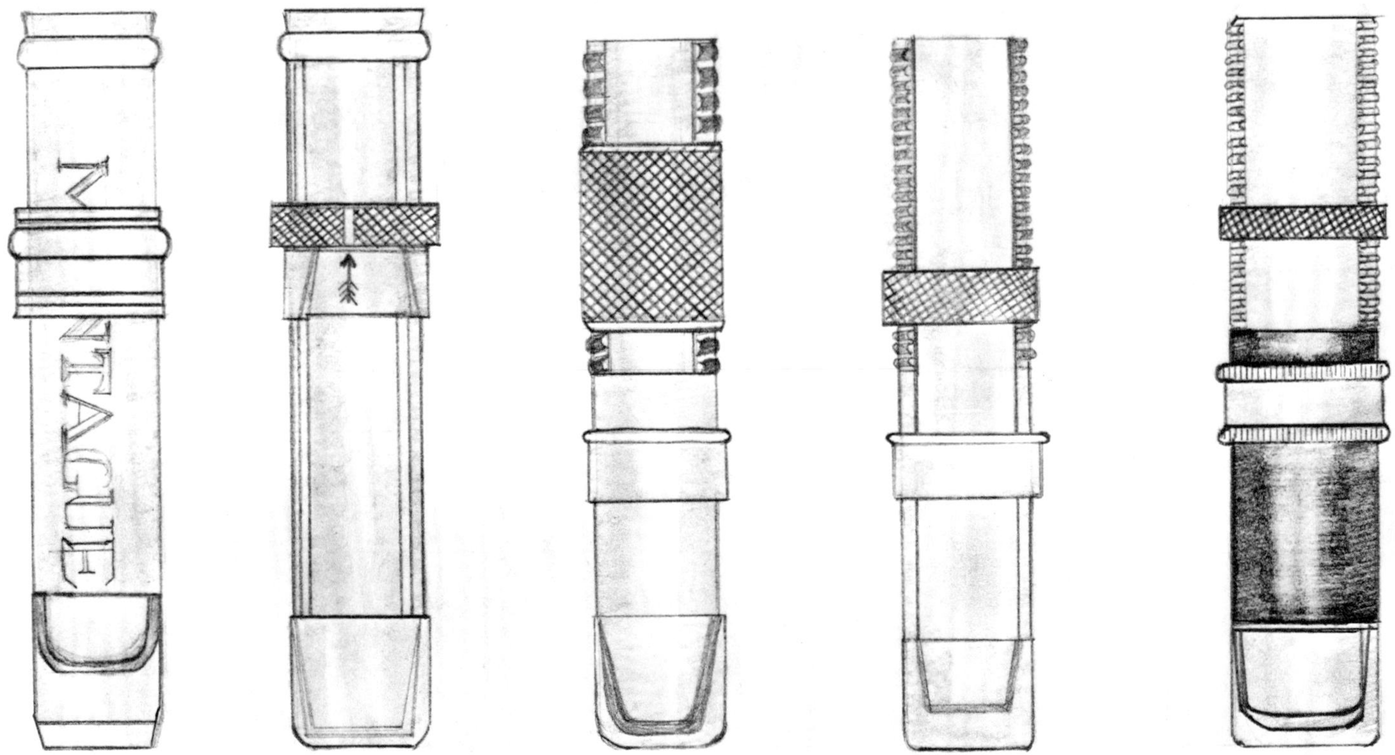

Sewell N. Dunton & Son, Inc.

Greenfield, Massachusetts 01301

4 Fiske Ave. Tel. 413-773-5453

MANUFACTURERS OF

Anglers' choice split-bamboo Rods

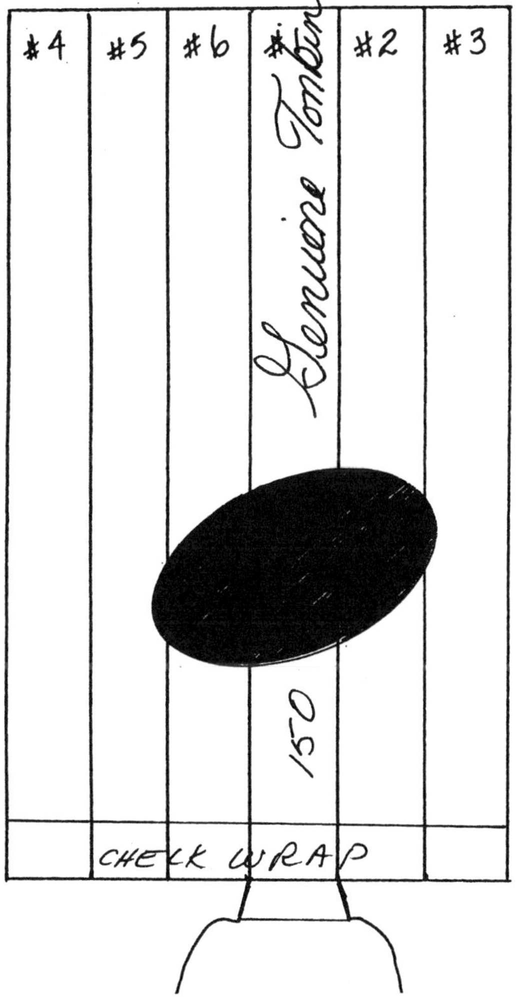

ANGLERS' CHOICE
SPLIT BAMBOO

ANGLERS' CHOICE
Sewell N Dunton & Son
Greenfield, Mass, U.S.A.

Phillipson Rod & Tackle Company

After World War II, Bill Phillipson attempted to buy the Goodwin Granger Co. from Granger's family. Phillipson knew that there would be an incredibly high demand for fishing rods among returning GI's, and believed it was imperative that production begin immediately. A decision could not be reached by the Granger family in a timely manner, so Phillipson began his own rod company. It has been said that Phillipson rods were *Granger look-alikes*. This simply is not true. There are similarities, to be sure, but the rods are more like cousins than brothers in both cosmetics and hardware.

Phillipson rods were assembled using a rescorcinal glue that left fine bluish lines between the strips at the joints. These are often criticized as being unattractive, but Phillipson felt they were an unavoidable side effect of using an absolutely superior glue. (The same glue was used to assemble plywood gliders in WW II). This is the easiest characteristic to spot in a Phillipson rod. The cane was treated in the same manner as Granger, and the nodes were arranged in the familiar 3 x 3 spacing. This was not a *copy cat* effort - it was simply an emulation of a proven and highly successful system – the system that Phillipson learned from Goodwin Granger while working as his foreman from 1925 until 1941. After Granger's death in 1931, Phillipson was completely in charge of production at the Granger plant.

Phillipson made a few improvements to his ferrules, though they were essentially the same as Granger. The slide portion of the Phillipson ferrule is shorter, and the rolled welt is less angular and appears to have two tiny steps. The ferrules, like Granger's, were not serrated and were never pinned.

Reelseats for Phillipson rods were of many, many types. Fortunately, there were only two basic themes for reelseats, and the many variations are the result of tiny changes that create a new style. Often the only change was the number of threads per inch used for the downlocking mechanism, or a new design for the threads. The two reelseats pictured on the flat map are the basic styles. The plastic seat was used on all models from 1946 to 1948, with lower grades using black plastic and higher grades using brown (butterscotch) plastic. The plastic seat was modified in about 1949 to use an aluminum threaded portion with the spacer remaining black plastic. After 1948, the black all plastic seat was always used on the lowest grade rods.

The *No–Rock* reelseat was said to have been patented, but no records have surfaced. This was the style used on higher quality rods after 1948. It may be seen in black, brown, or green anodizing, and was sold through the catalog with no anodizing. Any rod that has a *natural* aluminum reelseat installed is one that is not *as original*, even though the reelseat is actually a Phillipson product and is an up-grade from the all plastic seat.

The thread used to wrap most Phillipson rods was nylon. Only the two highest grades and some special models used silk. Nylon thread was less expensive, and was the current *modern technology*, as Nylon was invented during WW II. In fact, it has been said that *Nylon* is an anagram for "Now You Look Out, Nippon!" These are the words that were rumored to have been spoken when the first fine Nylon filaments were drawn. They were used to replace human hair for the *cross-hairs* in the bomb sights used on the B-25 and B-29 aircraft in WW II. Whatever the reason, Phillipson elected to use nylon on his lower grades from the beginning, and later used it for all of his rods. Nylon thread is widely available, and may be ordered from any of the suppliers listed in the appendix. Here are the color names and color numbers from Gudebrod:

Black = 001, Yellow = 6778, Lime = 105. (this color was discontinued by the manufacturer in 1951, and so beginning in 1952, the wraps for the Pacemaker and the new Smuggler changed to Leaf Green tipped black) Leaf Green = 6779

Turf Tan was a special color made for Phillipson for use on the *Peerless* series. In Nylon, use #290, though it is a bit light. In silk use #832, called *beige*. It is closer to the real Turf-Tan.

Markings on Phillipson rods are straightforward. The maker, model name, length of the rod and recommended line size are all written on two flats. Occasionally, a small number is written near the winding check wrap. This is the employee number of the final inspector and was added after the rod was wrapped and before it went to varnish. Phillipson rods do not have hookkeepers. They do have a NS winding check that is the same small, plain check used by Granger and several other makers.

An unusual feature of Phillipson rods is the Elliptical Full Wells Grip, also referred to as the *Hammer-Handle* grip. It was designed by Phillipson while at the Granger Company, and was used on the two-piece Granger rods. The Hammer-Handle was used by Phillipson on his highest grades, the Paramount and Premium. Later it was used on the early Peerless rods. These early Peerless rods were impregnated, and the color of the cane resembled mahogany.

The wraps are actually black, tipped white and black, but they appear to be black and gold due to yellowed varnish. These are rarely found, and are sometimes thought to be the Premium model due to the wrap colors. They are usually marked "Impregnated" on the rodshaft, and sometimes do not have the markings of a Peerless rod. Oxidized ferrules are another clue, but these were also used on the early Premium models, so the markings and the color of the cane are the best clues. Phillipson's Peerless series was in production under ownership of the 3M Company as late as 1974, and was the last maker of bamboo rods at affordable prices.

Phillipson Elliptical Full Wells

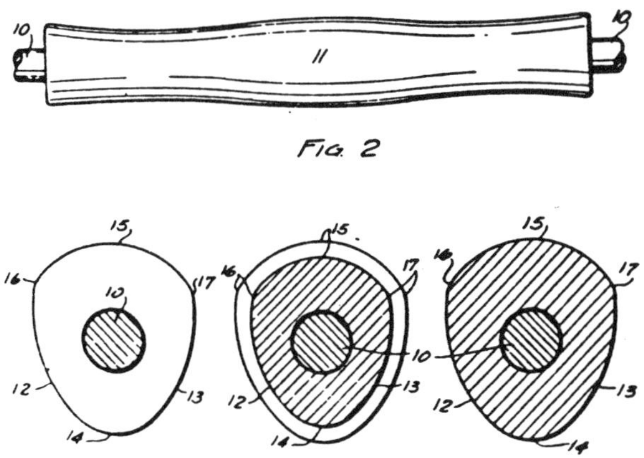

Phillipson WORLD'S FINEST ®

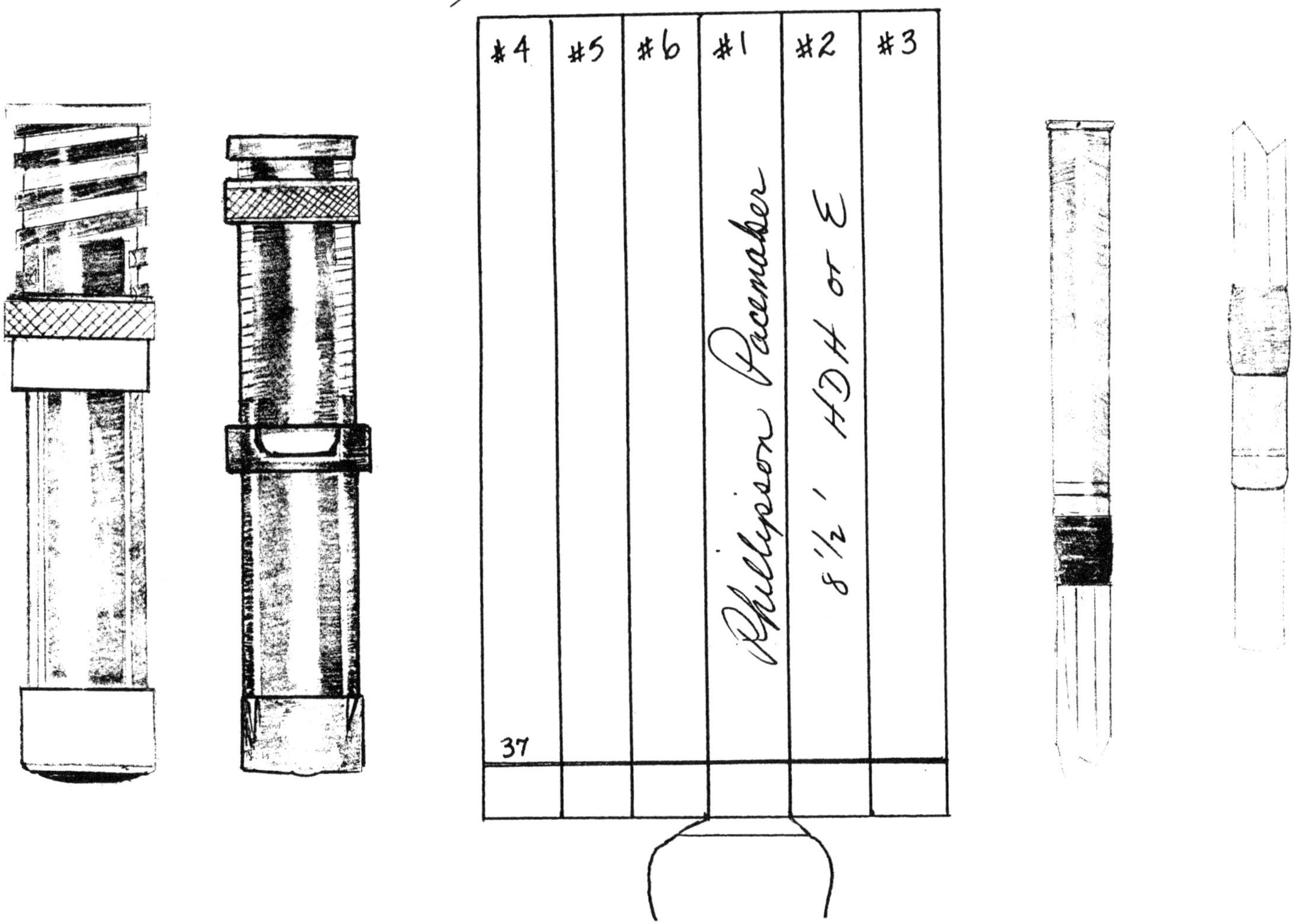

Phillipson Cork Grip Measurements

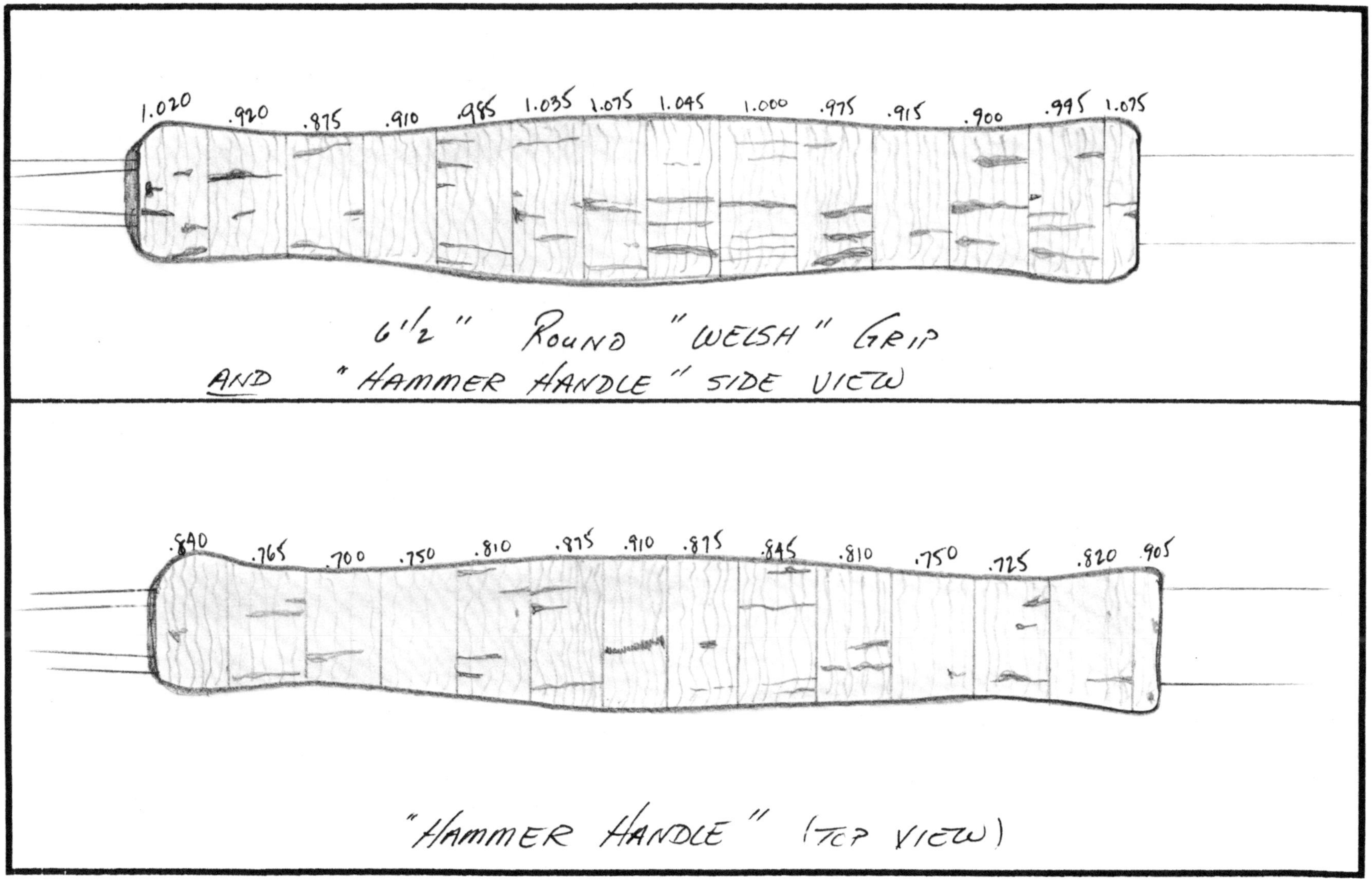

Phillipson Rod & Tackle Co. **Hardware Specs**

Year	Model	$	Reelseat	Grip	Stripper	Snake	TT	Ferrules
1946	Pacemaker	25.00	BK Plastic, SDL	Welsh	P, BK	P, BTS	P, BTS	NS
	Pow'r Pak't	37.50	BN Plastic, AL SDL	Welsh	P, BK	P, BTS	P, BTS	NS
	Paramount	50.00	BN Plastic, AL SDL	Hammer	P, C	P, TS	P, C	NS
	Premium	75.00	BN Plastic, AL SDL	Hammer	P, C	P, TS	P, C	NS, OX
1949	Paragon	19.75	BK Plastic, SDL	Welsh	BK, WS	P, BTS	P, BTS	NS
	Pacemaker	25.00	BK Plastic, AL Thread SDL	Welsh	BK, WS	P, BTS	P, BTS	NS
	Pow'r Pak't	37.50	BN Plastic, AL Thread SDL	Welsh	BK, WS	P, TS	P, TS	NS
	Paramount	50.00	BN Plastic, AL Thread SDL	Hammer	P, C	P, TS	P, C	NS
	Premium	75.00	BN Plastic, AL Thread SDL	Hammer	P, C	P, TS	P, C	NS
1952	Pacemaker	25.00	BK Plastic, AL Thread SDL	Welsh	BK, WS	P, TS	P, TS	NS
	Smuggler	25.00	GN, AN, AL No-Rock SDL	Welsh	BK, WS	P, TS	P, TS	NS
	Pow'r Pak't	37.50	BN, AN, AL No-Rock SDL	Welsh	P, C	SS	P, C	NS
	Paramount	50.00	BN, AN, AL No-Rock SDL	Hammer	P, C	SS	P, C	NS
	Premium	75.00	BK, AN, AL No-Rock SDL	Hammer	P, C	SS	P, C	NS, OX
1953	Preferred	42.50	BN, AN, AL No-Rock SDL	Hammer	P, C	SS	P, C	NS
	Peerless	75.00	BK, AN, AL No-Rock SDL	Hammer	P, C	SS	P, C	NS
1964	Peerless	80.00	BN, AN, AL No-Rock SDL	Torpedo	P, C	SS	P, C	NS
	Peerless "5"		BN, AN, 2 Rings on Cork	Cigar	P, C	SS	P, C	NS
1972	Dry Fly Spec.		BN, AN, AL No-Rock SDL	Torpedo	P, C	SS	P, C	NS
	Alternate		BN, AN, 2 Rings on Cork	Cigar	Snake	SS	P, C	NS

Phillipson Special Models/Trade Rods — Hardware Specs

Year	Model	Reelseat	Grip	Stripper	Snake	TT	Ferrules
1947	Haywood Zephyr	SDL, GLD/AN, Brown Comp.	Hammer	PBK	PSS	SS	SS
	Haywood Zephyr	Same, w/2" Ext Butt					
	Ed M. Hunter	SDL, BK/AN No-Rock	Welch	PBK	PB	PB	NS, K
		W/ Daisy Butt Cap					
	A&F Firehole	BRZ, AN, No-Rock	Torpedo	PC	PC	PC	NS
	A&F Yellowstone	BRZ, AN, No-Rock	Torpedo	PC	PC	PC	NS
	A&F Teton	BRZ, AN, No-Rock	Torpedo	PC	PC	PC	NS
1953	Colorado Tackle Co.						
	Topflyte ($16.50)	AL, SDL	Welsh	TS	TS	TS	NS, RW

Phillipson Rod & Tackle Co. — Wrap Specs

Era	Model	Thread	Wraps	Tipping	Notes on Special Wraps
1946	Pacemaker	Nylon	Lime Green	Yellow	None
	Pow'r Pak't	Nylon	Yellow	Black	None
	Paramount	Silk	B&W Jasper #X	Black/Gold	Tipped Black, Gold Black
1948	Premium	Silk	Black #1	Black/Gold	Tipped Gold, Black
1949	Paragon	Nylon	Beige #290	None	3 narrow bands above Check Wrap
to '51	All Other Models as Above				
1952	Pacemaker	Nylon	Leaf Green	Black	None
	Smuggler	Nylon	Leaf Green	Black	None
	Pow'e Pak't	Nylon	Yellow	Black	None
	Paramount	Silk	B&W Jasper #X	Black/Gold #7	Tipped Black, Gold, Black
1954	Premium	Silk	Black #1	Black/Gold #7	Tipped Black, Gold, Black
1953	Preferred	Silk	Gold #7	Black #1	Impregnated, only wraps varnished
to '60	Peerless	Silk	Black #1	White #2	Tipped Black, White, Black - impregnated
1961	Peerless	Nylon	Turf Tan	Yellow	Open Spiral, Yellow Tipping
Special Models:					
1947	Haywood Zephyr	Silk	Gold #7	Black #1	
	Ed M. Hunter	Silk	Black #1	None	
1953	Top-Flyte	Nylon	Red #10	None	

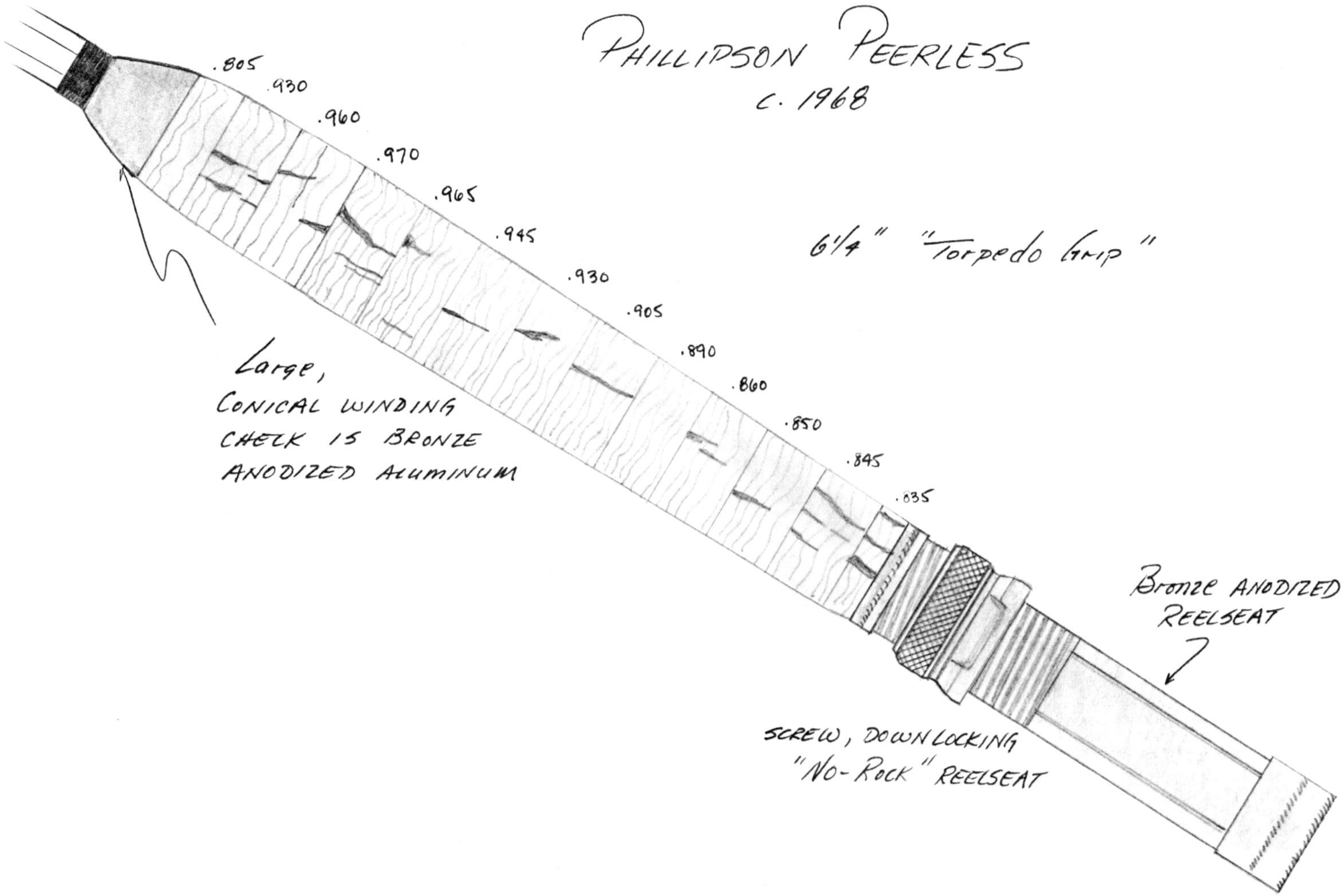

PHILLIPSON PEERLESS
c. 1968
6 1/4" "Torpedo Grip"
.805
.930
.960
.970
.965
.945
.930
.905
.890
.860
.850
.845
.835
Large, CONICAL WINDING CHECK IS BRONZE ANODIZED ALUMINUM
Bronze ANODIZED REELSEAT
SCREW, DOWNLOCKING "NO-ROCK" REELSEAT

William Shakespeare Company

Yes, there really was a William Shakespeare, Jr. He was born in 1869, the son of General William Shakespeare of Civil War fame. In addition to founding the tackle firm that bears his name, William Shakespeare, Jr. also served as mayor of Kalamazoo, Michigan. from 1933 to 1937. After his political career ended, Shakespeare moved to Melbourne, Florida where he established a radio station and the Federal Home Savings Association. Mr. Shakespeare died in Melbourne in 1949 at the age of 80.

The William Shakespeare Company began as a small one-man operation. Shakespeare was an ardent fisherman and wanted to design and make a reel that did not require manipulation of the line with a finger to make it wind evenly onto the reel. The first patent for this mechanism was granted in 1896, and the business of manufacturing the reels was initiated the following year. In 1905, the company was incorporated and moved to the Hanselman building at the corner of Rose and Water Streets in downtown Kalamazoo. By 1913, a new building had been erected as the home of the company, which still stands today. This huge facility covers a very large city block in Kalamazoo.

In an effort to provide a more diversified company, Shakespeare formed a new company called Shakepeare Products in 1921. Through two world wars the Shakespeare companies manufactured vital military equipment.

By 1924, Shakespeare offered four models of fly rods. They were of upper middle to lower middle grades and sold for prices of $12.00 to $24.00. All of these were known as *Shakespeare Ideal Rods*. Each model was designated by number. These rods appear to have been made by Montague, as they have ferrules and other hardware that resembles Montagues of the era. It seems that Shakespeare never made bamboo rods in its factory and contracted with several different firms throughout its history to supply the rods sold under the Shakespeare label. Early on, Shakespeare's rod specs had enough differences from the rods sold under the manufacturers name to make them seem like the work of an independent maker. Later, they were exact copies of the maker's work with only the Shakepeare decal and a slightly modified reelseat to differentiate them.

In the period from about 1926 to 1934, Shakespeare began to use model names as well as numbers to designate the rod grade. The letter "B" followed most rod model numbers, and seems to have no significance as to action since *light calibre* and *medium calibre* rods had the same letter as a suffix. The higher grade rods (from the Favorite and higher) were made by Heddon, and those from the Trout River down were by Montague. The Heddon models are the usual Tempered Cane finish used on all Heddon fly rods, and the Montagues used natural colored cane.

1935 marked a year of changes for Shakespeare rods. The first change was in the model number. Following the model number is a letter that designates the action of the rod. The letters and actions they indicate are:

T = Trout weight
B = Bass weight
L = Light calibre (Light Dry Fly)
M = Medium calibre (All-Around action)
S = Select calibre (Dry Fly action)

Some models were made in one action only, and did not have a letter in the model number.

The next change in the rod line for 1935 was the addition of South Bend as a supplier of rods. South Bend supplied the top-grade Superba Double-Built in two calibres, and single-built rods in several grades. Other suppliers remained Heddon and Montague. Just prior to the beginning of WW II, Shakespeare dramatically cut the rod line. Only eleven models were made. South Bend was the major manufacturer, providing eight of the models. The remaining three (the lowest grades) were Montague products.

The reelseats were similar to those used by South Bend, but were all of uplocking design. Most of the rods had what was called a Snap-Lock reelseat. This aluminum seat was anodized in several colors. The seat looks like the South Bend Lite-Lock seat, but rather than having an internal thread for adjustments, the Snap-Lock has an internal spring. Thus to operate the reelseat, one simply pulled the cap down, inserted the reelfoot, and released the cap. A sound and simple system - as long as the spring is not overstressed!

After WW II, the bamboo rod line was again reduced - this time to only three models, all of middle to lower grades and made by South Bend. In 1947, Shakespeare, the company that had never made rods, changed fishing tackle with the introduction of a brand new rod-making material: Fiberglass. The developement of the glass rod began as the result of a broken fly rod tip. The rod belonged to Dr. Arthur M. Howald of Toledo, a dyed-in-the-wool angler and technical director for the Plaskon Division of Libby-Owens Ford Glass Co. In his wartime experiments he discovered many uses for the Plaskon-treated fiber glass. He tried using it to make a new tip for his flyrod - it worked! Howald then made plans to make an entire rod of fiber glass. The entire process of engineering and manufacturing the Howald *Glastik* rod was taken over by Shakespeare, and the *Wonderod* was born. In today's collectible market, it might be interesting to note that the original 1947 Wonderods (two models) sold for $60.00, and the bamboo rods sold by Shakespeare in the same catalog went for $16.00 to $26.50.

On the flat map are three Shakespeare decals. The dark decal on the right is a logo that was used throughout Shakespeare history and is rarely seen on bamboo rods. The two on the left are the commonly seen rod decals. The older one is at the top. This is the pre-WW II decal. The lower decal was used after WW II. Both decals are olive-green ovals with a gold foil banner bearing the word Shakespeare in its familiar script style lettering. The remainder of the lettering is plain black block style. The decal is placed so that the lettering reads *away from* the grip toward the ferrule. On some older bait rods, the decal is rotated 90° to the rodshaft.

Shakespeare Company
KALAMAZOO, MICHIGAN., U.S.A.

HONOR BUILT
Shakespeare
HONOR SOLD

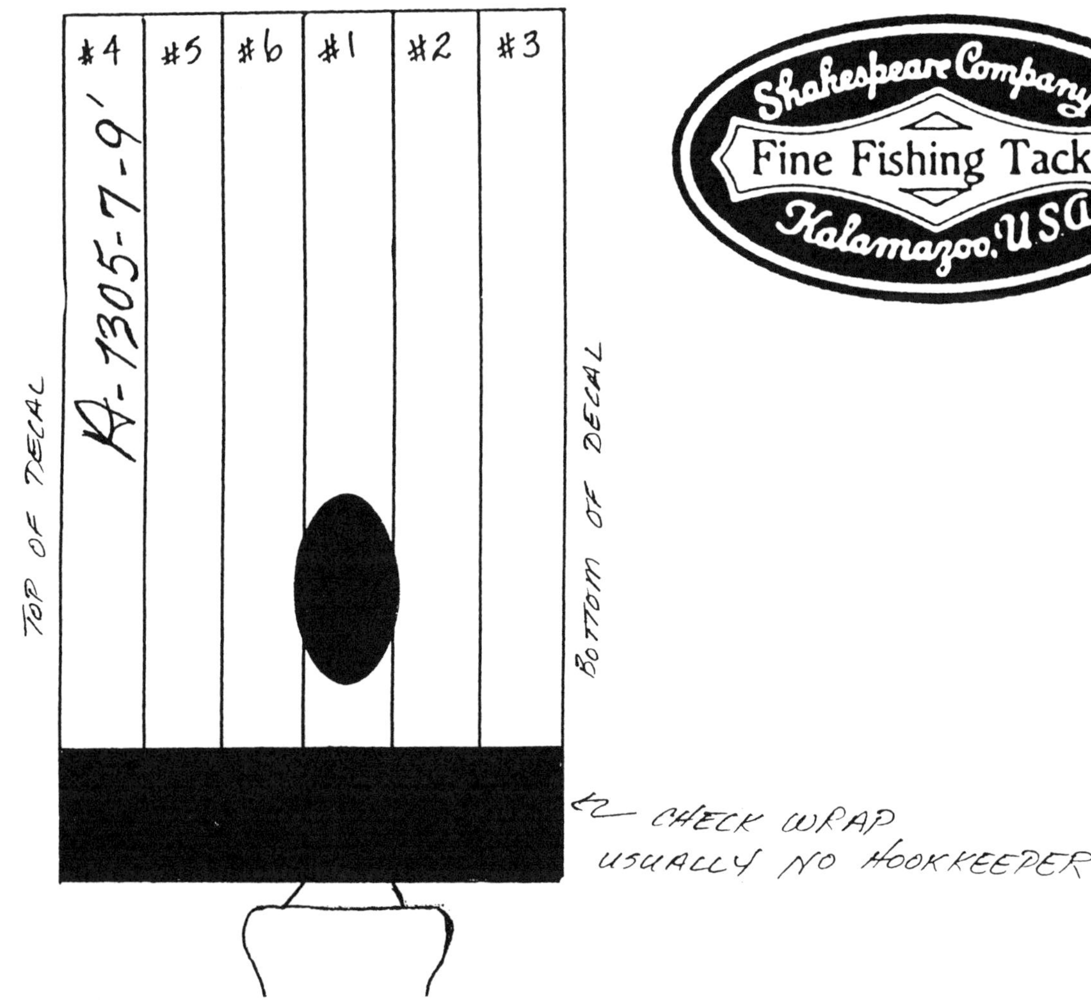

Shakespeare Company
Fine Fishing Tackle
Kalamazoo, U.S.A.

HONOR BUILT - HONOR SOLD
Shakespeare
MADE IN U.S.A.

Model	$	Action	Cane	Reelseat	Grip	Stripper	Snake	TT	Ferrule
3310	12.00	Bass	Natural	NPB, SB	6" Full Wells	A	SS	SS	NPB, RW, SH
3318	12.00	Trout	Natural	NPB, SB	6" Full Wells	A	SS	SS	NPB, RW
3328	18.00	Trout	Natural	NPB, SB	6" Full Wells	A	Bronze	SS	NPB, RW, SH
3338	24.00	Trout	Natural	NS, SB	6" Full Wells	A	Bronze	SS	NS, RW, SH

Model	Main Color	Tipping Color	Intermediates	Notes
3310	Black #1	Red #10	Red #10	Closed HK, Intermediates widely spaced
3318	Gold #7	Black #1	Black #1	Closed HK, Intermediates widely spaced
3328	Green #24	Gold #7	Gold #7	Closed HK, Intermediates widely spaced
3338	Gold #7	Red 310	Red #10	Closed HK, Intermediates closely spaced

Shakespeare Company — Hardware Specs — 1930

Model	Name	$	Reelseat	Grip	Stripper	Snake	TT	Ferrules
1338B	DeLuxe (L)	30.00	NSB,C&R	6½" Full Wells	Agate	HS	PS	NS, HW, S
1315B	DeLuxe (M)	30.00	NSB, C&R	7½" Full Wells	Agate	HS	PS	NS, HW, S
1328B	Intrinsic (L)	20.00	NS, C&R	6½" Full Wells	Agate	HS	PS	NS, W, S
1311B	Intrinsic (M)	20.00	NS, C&R	7½" Full Wells	Agate	HS	PS	NS, W, S
1309B	Tillamook	20.00	NS, C&R	7½" Rev. HW	Agate	HS	PS	NS, W, S
1330B	Rangely	20.00	NS, C&R	6½" Rev. HW	Agate	HS	PS	NS, W, S
1300B	Favorite	12.50	NS, C&R	6½" Full Wells	SS	SS	SS	NS, HW
1310B	Perfect	12.50	NS, C&R	6½" Full Wells	SS	SS	SS	NS, HW
1312	Trout River	7.50	NPB, C&R	6" Rev. HW	SS	SS	SS	NS, RW, SH
1305	Manistee	5.00	NPB, C&R	6" Rev. HW	SS	SS	SS	NS, RW, SH

Shakespeare Company — Wrap Specs — 1930

Model	Name	Main Color	Tipping	Intermediates	Notes
1338B	DeLuxe (L)	Yellow #3	Black #1	N/A	Hookkeeper
1315B	DeLuxe (M)	Tan #27	Purple #16	N/A	Hookkeeper
1328B	Intrinsic (L)	Green #24	Black #1	N/A	Hookkeeper
1311B	Intrinsic (M)	Red #10	Black #1	N/A	Hookkeeper
1309B	Tillamook	Green #20	Black #1	N/A	Hookkeeper
1330B	Rangely	Green #20	Black #1	N/A	Hookkeeper
1300B	Favorite	Green #24	Red #10	N/A	Hookkeeper
1310B	Perfect	Green #24	Red #10	N/A	Hookkeeper
1312	Trout River	Red #1	Black #1	N/A	Hookkeeper
1305	Manistee	Green #24	N/A	Green #24	Hookkeeper

Shakespeare Company   Hardware Specs   1931

Model	Name	$	Reelseat	Grip	Stripper	Snake	TT	Ferrules
1340B	Superba	40.00	SDL, AL & NSB	6½" Rev. HW	Agate	HS	PS	NSB, HW, S
1341B	Superba	45.00	Custom Order Only – Customer's Specs					
1240B	Gypsy 7½' - 2/2	40.00	SDL, NS, Walnut	6½" Rev. HW	Agate	HS	PS	NS, HW, S
1233B	Diana 6½' - 2/2	15.00	NS, C&R	6½" Rev. HW	Agate	HS	PS	NS, HW, S
1338B	DeLuxe (L)	30.00	BNS, C&R	6½" Full Wells	Agate	HS	PS	NSB, HW, S
1315B	DeLuxe (M)	30.00	BNS, C&R	6½" Full Wells	Agate	HS	PS	NSB, HW, S
1332B	Superior (L)	25.00	NS, SDL	6½" Rev. HW	Agate	HS	PS	NS, RW, SH
1314B	Superior (M)	25.00	NS, SDL	6½" Rev. HW	Agate	HS	PS	NS, RW, SH
1328B	Intrinsic (L)	20.00	NS, C&R	6½" Full Wells	Crystal	HS	HS	NS, RW, S
1311B	Intrinsic (M)	20.00	NS, C&R	6½" Full Wells	Crystal	HS	HS	NS, RW, S
1309B	Tilamook (H)	20.00	NS, C&R	7½" Rev. HW	Agate	HS	HS	NS, RW
1330B	Rangely (L)	20.00	NS, C&R	6½" Rev. HW	Agate	HS	HS	NS, RW
1300B	Favorite	10.00	NS, C&R	6½" Full Wells	Crystal	HS	HS	NS, RW
1313	Trout River	7.00	NPB, C&R	6" Cigar	Black, S	HS	HS	NS, RW, SH
1303B	Triumph	15.00	NS, C&R	6½" Peckerhead	Crystal	HS	HS	NS, RW, S

Shakespeare Company Wrap Specs 1930

Model	Name	Main Wraps	Tipping	Intermediates	Notes
1341B	Superba Spec.	Custom Order Only – Customer's Specs			
1340B	Superba	Green #24	Yellow #3/Red #10	N/A	Hookkeeper
1240B	Gypsy	Yellow #3	Blue #17	N/A	Hookkeeper
1233B	Diana	Yellow #3	Blue #17	N/A	Hookkeeper
1338B	DeLuxe	Yellow #3	Black #1	N/A	Hookkeeper
1315B	DeLuxe	Yellow #3	Puple #16	N/A	Hookkeeper
1332B	Superior	Black #1	Yellow #3/Red #10	N/A	Hookkeeper
1314B	Superior	Black #1	Yellow #3/Red #10	N/A	Hookkeeper
1328B	Intrinsic	Green #22	N/A	N/A	Hookkeeper
1311B	Intrinsic	Red #10	N/A	N/A	Hookkeeper
1309B	Tilamook	Green #24	Black #1	N/A	Hookkeeper
1330B	Rangely	Green #24	Black #1	N/A	Hookkeeper
1300B	Favorite	Green #24	Red #10	N/A	Hookkeeper
1313	Trout River	Red #10	Black #1	N/A	Hookkeeper
1303B	Triumph	Green #24	N/A	N/A	Hookkeeper

Shakespeare Company   Hardware Specs   1936

Model	Name	$	Cane	Reelseat	Grip	Stripper	Snake	TT	Ferrules
1344	Superba	50.00	B	BNS, UL	Rev. HW	A	HS	HS	BNS, HW, S, w/o
1343	Superba	50.00	B	BNS, UL	Rev. HW	A	HS	HS	BNS, HW, S, w/o
1342	Superba	40.00	B	BNS, UL	Rev. HW	A	HS	HS	BNS, HW, S, w/o
1315T	DeLuxe	30.00	BT	UL, Black	Heddon HW	CY	HS	PS	GMNS, HW, S, w/o
1315B	DeLuxe	30.00	BT	UL, Black	Heddon HW	CY	HS	PS	BNS, HW, S, w/o
1314T	Superior	25.00	BT	UL, Black	Heddon HW	Ag	HS	PS	BNS, HW, S, w/o
1314B	Superior	25.00	BT	UL, Black	Heddon HW	Ag	HS	PS	BNS, HW, S, w/o
1312T	Supreme	20.00	F	AL, UL, Black	SB Half Wells	Ag	HS	HS	NS, RW, w/o
1312B	Supreme	20.00	F	AL, UL, Black	SB Half Wells	Ag	HS	HS	NS, RW, w/o
1309	Tilamook (H)	16.00	F	AL, UL, Black	Rev. HW	A	HS	HS	NS, RW, SH
1311L	Intrinsic (L)	18.00	LB	AL, UL, Black	Rev. HW	A	HS	HS	BNS, HW, S, w/o
1311S	Intrinsic (M)	18.00	LB	AL, UL, Black	Rev. HW	A	HS	HS	BNS, HW, S, w/o
1311M	Intrinsic	18.00	LB	AL, UL, Black	Rev. HW	A	HS	HS	BNS, HW, S, w/o
1307	Premier	15.00	BT	Heddon UL	Heddon HW	Ag	HS	HS	NS, RW, w/o
1306	Niagra	15.00	F	AL, DL, HR	Straight	Ag	HS	HS	CDNS, RW, w/o
1304	Paragon	12.00	N	SDL, AL	Half Wells	Ag	HS	HS	BNS, HW, S, w/o
1223	Diana	10.00	B	NS, SB	Comp. Cork	CY	HS	HS	NS, RW, S, w/o
1350	Texas Bug Rod	15.00	LB	AL, UL Bake.	Half Wells	A	HS	HS	NS, RW, S
1300	Favorite	10.00	MB	CPB, SDL	Half Wells	Ag	HS	HS	NS, RW, w/o
1303L	Triumph	10.00	LB	SDL, Bakelite	Cigar	CY	HS	HS	NS, RW, SH
1303S	Triumph	10.00	LB	SDL, Bakelite	Cigar	CY	HS	HS	NS, RW, Straight
1303M	Triumph	10.00	LB	SDL, Bakelite	Cigar	CY	HS	HS	NS, RW, SH

Shakespeare Company **Hardware Specs** **1936**

Model	Name	$	Cane	Reelseat	Grip	Stripper	Snake	TT	Ferrules
1363	AuSable	7.70	F	NPB, SDL	Half Wells	Ag	HS	HS	NPB, RW, SH
1362	Spring Brook	6.60	F	NPB, SDL	Half Wells	Ag	HS	HS	NPB, RW, SH
1358	Service	3.30	MB	NPB, SDL	Half Wells	SN	HS	HS	NPB, RW, SH
1361	Pine River	5.50	B/LB	NPB, SDL	Straight	SN	HS	HS	NPB, RW, SH
1360	Silver Creek	4.60	N	NPB, SB	HW, Comp.	Ring	HS	HS	NPB, RW, SH
1357	Marvel	3.00	N	NPB, SB	HW, Comp.	SN	HS	HS	NPB, RW, SH

Shakespeare Company   Wrap Specs   1936

Model Name	Main Wraps	Tipping	Intermediates	Notes
Superba	Green #24	Yellow #3/ Red #10	Groups	Groups of 5 - G, Y, R, Y, G
DeLuxe	Maroon #12	Yellow #3	None	Open HK wrap
Superior	Black #1	Yellow #3	None	Open HK wrap
Supreme	Blue #17	White #2	None	Closed HK wrap
Tilamook	Black #1	Green #19	None	Closed HK wrap
Intrinsic	Yellow #3	Black #1	None	Closed HK wrap
Premier	Crimson #10	N/A	None	Closed HK wrap
Niagra	Green #24	Yellow #3	None	Closed HK wrap
Paragon	Brown #27	Brown #28	None	Closed HK wrap
Diana	Green #23	Red #10	None	Closed HK wrap
Texas Bug Rod	Black #1	Orange #8	None	Closed HK wrap
Favorite	Green #24	Red #10	Red #10/Green #24	Alternating colors
Triumph	Green #23	Red #10	None	Closed HK wrap
AuSable	Purple #16	Gold #7	Purple #16/Gold #7	Alternating colors
Spring Brook	Orange #8	Black #1	Black #1	Closed HK wrap
Service	Green #24	None	Green #24	Closed HK wrap
Pine River	Purple #16	Red #10	None	Closed HK wrap
Silver Creek	Green #24	Orange #8	None	Closed HK wrap
Marvel	Red #10	Green #23	None	Tipped at HK & ferrules only

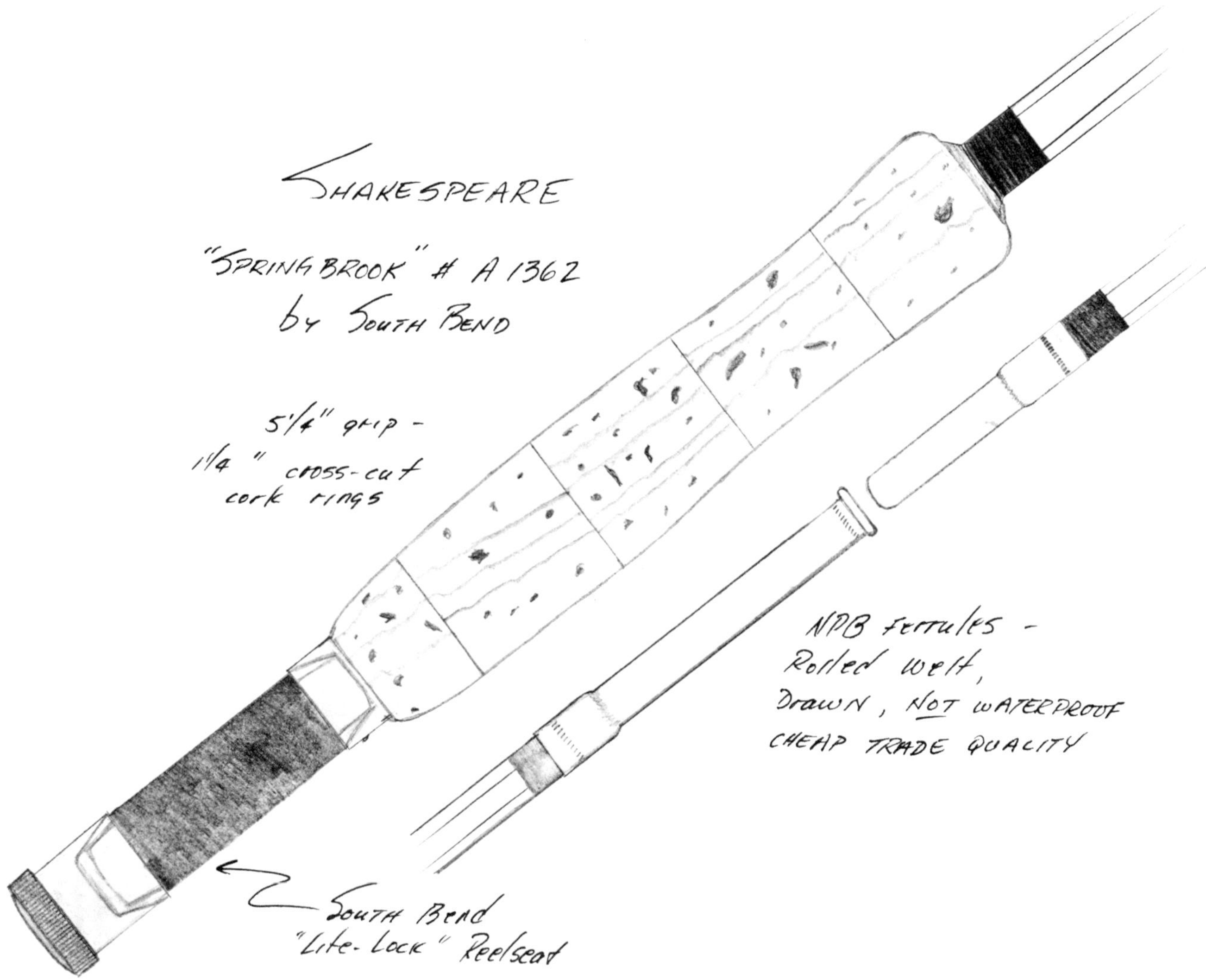
SHAKESPEARE
"SPRINGBROOK" # A 1362
by SOUTH BEND
5¼" grip -
1/4" cross-cut
cork rings
NPB ferrules -
Rolled welt,
Drawn, NOT WATERPROOF
CHEAP TRADE QUALITY
SOUTH BEND
"Lite-Lock" Reelseat

Shakespeare Company Hardware Specs 1942

Model	Name	$	Cane	Reelseat	Grip	Stripper	Snake	TT	Ferrules
1308	Premier	18.00	F	Snaplock, Blk	6½" Half Wells	Ag	HS	HS	NPB, BLK, RW, S
1233	Diana	14.25	DB	Snaplock, Grn	6½" Half Wells	Ag	HS	HS	NS, RW, S
1363	AuSable	11.50	F	Snaplock, Grn	6½" Half Wells	Ag	HS	HS	NPB, BLK, RW, SH
1360	Silver Creek	10.25	MB	Snaplock, Blk	6½" Full Wells	HS	HS	HS	NPB, BLK, RW, SH
1312	Tru-Art	22.50	F	Snaplock, Blk	6½" Full Wells	SS	HS	HS	NS, RW
1307	Premier	18.00	F	Snaplock, Blk	6½" Half Wells	Ag	HS	HS	BNS, RW
1234	Reynard	22.50	BT	DSL, Tenite	7" Str. Cigar	Ag	HS	HS	NS, RW
1300	Favorite	13.75	F	Snaplock, Blk	7" Str. Cigar	Ag	HS	HS	BNS, RW
1315	DeLuxe	34.00	F	S, UL, NS/AL	7¼" Full Wells	A	PC	PC	GMNS, HW, S,
1314	Superior	29.50	F	S, UL, NS/AL	7¼" Full Wells	A	HS	PC	GMNS, RW
1362	Spring Brook	9.75	N	SDL, Tenite	5½" Half Wells	Ag	HS	HS	CPB, RW, SH, K
1358	Service	6.25	N	CP, Screwlock	6" Straight	BS	HS	HS	NPB, RW, SH

Shakespeare Company **Wrap Specs** **1942**

Model Name	Main Wraps	Tipping	Intermediates	Notes
Premier	Red #10	Yellow #3	N/A	Hookkeeper
Diana	Green #23	Red #10	N/A	Hookkeeper
AuSable	Green #23	White #2	N/A	Hookkeeper
Silver Creek	B&O Jasper	N/A	N/A	Hookkeeper
Tru-Art	Blue #17	White #2	N/A	Hookkeeper
Premier	Red #11	Yellow #3	N/A	Hookkeeper
Reynard	Orange #8	Black #1	N/A	Hookkeeper
Favorite	Blue #17	Orange #8	N/A	Hookkeeper
DeLuxe	Maroon #12	Yellow #3	N/A	Hookkeeper
Superior	Black #1	Yellow #3	N/A	Hookkeeper
Spring Brook	B&W Jasper	N/A	N/A	Hookkeeper

Model	Name	$	Cane	Reelseat	Grip	Stripper	Snake	TT	Ferrules
A1305	Triumph	26.75	F	Snaplock, Grn.	SB Half Wells	G	HS	HS	CPNS, RW
A1305B	Triumph	26.65	F	Snaplock, Grn.	SB Half Wells	G	HS	HS	CPNS, RW
A1305T	Triumph	26.75	F	Snaplock, Grn.	SB Half Wells	G	HS	HS	CPNS, RW
A1362	Spring Brook	18.25	N	NPB, SDL	Rev. Half Wells	Ag	HS	HS	CPB, RW, SH
A1362B	Spring Brook	18.25	N	NPB, SDL	Rev. Half Wells	Ag	HS	HS	CPB, RW, SH
A1362T	Spring Brook	18.25	N	NPB, SDL	Rev. Half Wells	Ag	HS	HS	CPB, RW, SH
1357	Marvel	13.75	LB	SB Lite-Lock	SB Half Wells	HS	HS	HS	NPB Straight

Model Name	Main Wraps	Tipping	Intermediates	Notes
Triumph	Green #23	White #2	N/A	Closed HK wrap
Spring Brook	B&W Jasper	None	N/A	Closed HK wrap
Marvel	Brown #27	Green #23	N/A	Closed HK wrap

South Bend Bait Company

South Bend rods of the 1920's and earlier were contracted through the Winchester Rod Co. Winchester Rods were made by Eustis W. Edwards. When Edwards ended his contract with Winchester in 1925, South Bend was faced with the problem of finding a new source for rods.

The solution was found when the Cross Rod Co. of Lynn, Massachusetts was purchased by South Bend in 1926. Not only did South Bend acquire the machinery of the Cross Rod Company, but also the services of Wes Jordan, one of the finest rodmakers in the history of the business. For reference, the Cross Rod and Tackle Company was founded in 1920 by William R. Forsyth and William Cross, with Wes Jordan as rodmaker. They were famous for their *Double-Built* rods, and had a wide selection of single-built rods as well.

Two completely different lines of rods were produced. South Bend Rods were priced to compete in the production rod business against makers such as Montague, Horrocks-Ibbotson and Union Hardware. The Cross Rods were promoted as a higher quality line to compete with Granger, Thomas, Edwards, and Heddon. South Bend continued to produce the basic Cross line of single and double-built rods, but gradually cut the number of models until only the most popular remained. The Cross rods disappeared from the South Bend lineup after WW II.

For purposes of dating South Bend rods, the first decal was used in 1926, when Jordan came to South Bend. It is an oval shaped yellow and orange decal. The next handy dating tool is the Comficient Grip. This was a ventilated half-Wells style grip with a Thumb Rest indentation. The grip was said to be *comfortable and efficient,* hence Comficient. There was a gold foil decal on the thumb rest indention when the rods came from the factory.

This grip was first used in about 1939, and was introduced at about the same time as the Lite-Lock reelseat. After WW II, the decal was changed to a yellow rectangle with the lettering in orange. A second rectangular decal on the rodshaft carries the model number, length of the rod and the recommended line size. A third, very late style of decal is sometimes seen. When fiberglass rods began to be produced, a smaller decal was devised due to the smaller diameter of the rodshaft. This late decal is a long, narrow rectangle that simply states "Made By / South Bend" in two lines. Most of these are very simple black lettering on a tan background. This decal was also used on factory repaired and refinished rods, and thus may be seen on rods that are obviously from an earlier period.

South Bend rod model numbers are a bit confusing, but there is logic to the system. Each rod was made in three actions; Wet Fly or Trout Action, Bass Action, and Dry Fly Action. The model number designates the rod's grade, or quality and the series number represents the type of action. For example, #159 is Trout Action, #59 is Bass Action, and #359 is Dry Fly Action. All of these Model 59 rods have the same wraps, hardware and other specifications - *only the taper is different.*

Usually, the models follow this format, with a 100-series being the lightest, or Trout Action, the 300-series being the medium-fast to fast Dry-Fly action, and the two-digit number being the stiffest and fastest, or Bass Action. The 2-piece models use a 200-series number (#260 and #290) and are also Dry-Fly action.

SOUTH BEND BAIT COMPANY

Manufacturers of SOUTH BEND Quality Tackle IND. U.S.A.

SOUTH BEND
QUALITY TACKLE
BAIT COMPANY

COMFICIENT
Comfortable-Efficient
GRIP

Look for This
Label On the
"Thumb Seat"

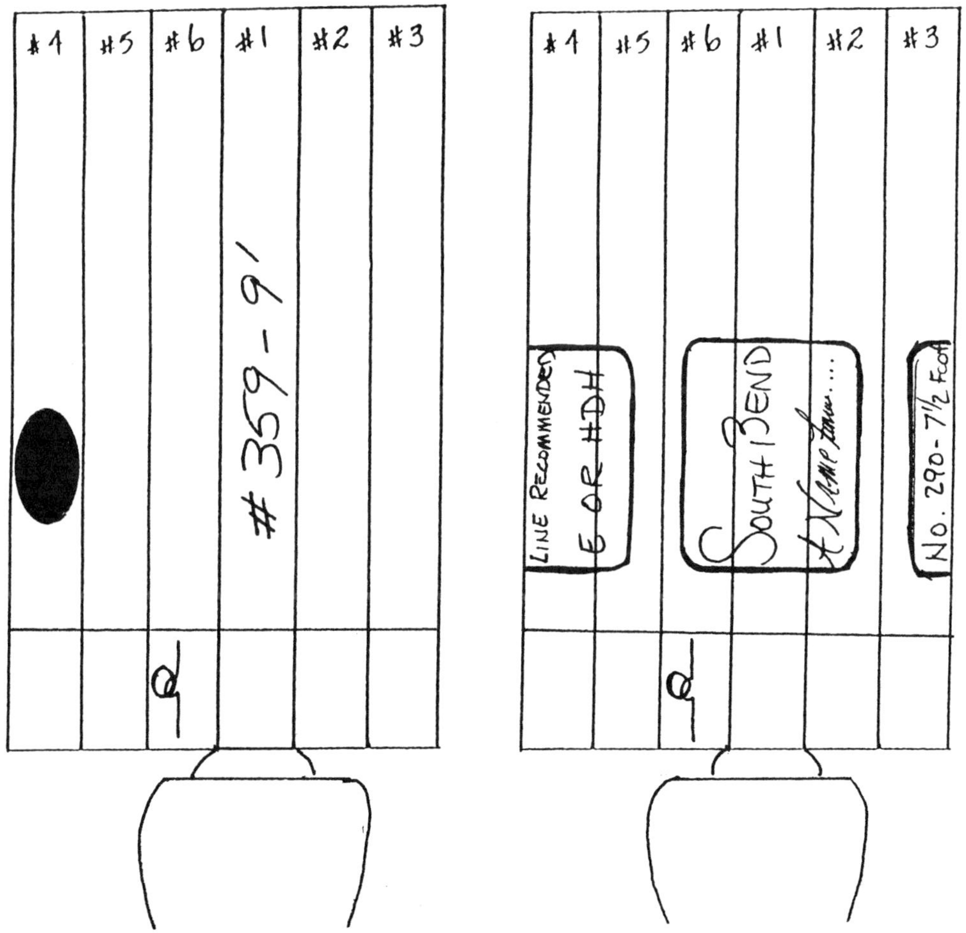

SOUTH BEND
A Name Famous in Fishing

No. 290 - 7 1/2 Foot

Line Recommended

E or HDH

Made by
SOUTH BEND

South Bend Bait Co. — Hardware Specs — 1920 - 30's

Era	Rod #	$	Cane	Reelseat	Grip	Stripper	Snake	TT	Ferrule
1925	30	30.00	N	NS, C&R	6½" Rev. HW	A	THSB	A	NS, HW, S
	20	18.00	N	NS, C&R	6½" Rev. HW	A	THSB	THSN	NS, HW,S
	12	12.00	N	NS, C&R	6" Cigar	THSN	THSN	THSN	NS, RW, S
1932	30/31	32.50	MB	Threadlock	6" Full Wells	A	SS	PC	SS, RW, S
	19/20	25.00	DB	Threadlock	7½" Rev. HW	A	SS	SS	NSCP, HW, S
	119/120	25.00	LB	Threadlock	7½" Rev. HW	A	SS	SS	SS, RW, S
	27/28	20.00	MB	Threadlock	7½" Rev. HW	Garnix	SS	SS	SS, RW, S
	12/13	16.00	N	Threadlock	6½" Cigar	Garnix	SS	SS	SS, RW, S
	26	12.50	LB	CP Screwlock	6" Cigar	SS	SS	SS	SS, RW, S
	66	5.00	DB	NPB, C&R	6" Full Wells	SS	SS	SS	NS, RW, S
	76	5.00	MB	NPB, C&R	6" Full Wells	SS	SS	SS	NPB, RW
	67	7.50	DB	NP Screwlock	6" Cigar	SS	SS	SS	NPB, RW
	77	7.50	MB	NP Screwlock	6" Cigar	SS	SS	SS	NPB, RW
	68	6.75	Silver	NPB, C&R	6" Full Wells	SS	SS	SS	NPB, RW
	25	20.00	N	Threadlock	6" Cigar	A	SS	SS	NSCP, HW, S
	29	7.50	DB	Threadlock	6" Cigar	Garnix	SS	SS	NS, HW, S

South Bend Bait Co. Wrap Specs 1920 - 30's

Era	Rod #	Main Wraps	Tipping	Intermediates	Notes	Signature
1925	30	Yellow #5	Red #10	Black #1	Standard Intermediates	Yes
	20	Yellow #5	Black #1	#5 & #1	Grouped #1, #5, #1	Yes
	12	Red #10	N/A	Red #10	Standard Intermediates	No
1932	30/31	Yellow #5	Red #10	Yellow #5	Standard Intermediates	Yes
	19/20	Yellow #5	Black #1	Yellow #5	Standard Intermediates	Yes
	119/120	Yellow #5	Black #1	Black #1	Standard Intermediates	Yes
	27/28	B&W Jasper	Orange #8	Orange #8	Standard Intermediates	Yes
	12/13	Red #10	Black #1	N/A	N/A	Yes
	26	Red #10	N/A	N/A	N/A	No
	66	B&O Jasper	N/A	N/A	N/A	No
	76	B&O Jasper	N/A	N/A	N/A	No
	67	B&O Jasper	Black #1	N/A	N/A	No
	77	B&O Jasper	Black #1	N/A	N/A	No
	68	Black #1	N/A	N/A	N/A	No
	25	B&O Jasper	Black #1	N/A	N/A	No
	29	B&O Jasper	N/A	N/A	N/A	No

South Bend Bait Co. Hardware Specs 1936

Rod #	$	Cane	Reelseat	Grip	Stripper	Snake	TT	Ferrule
30	35.00	B	Threadlock	6" Full Wells	A	THS	THS	NSCP, HW, S
31	35.00	B	Threadlock	7½" Full Wells	A	THS	THS	NSCP, HW, S
119	25.00	MB	Threadlock	6" Rev. HW	A	THS	THS	NSCP, HW, S
120	25.00	MB	Threadlock	7½" Rev. HW	A	THS	THS	NSCP, HW, S
27	20.00	MB	Threadlock	6" Rev. HW	A	THS	THS	NSCP, RW, S
28	20.00	MB	Threadlock	7½" Rev. HW	A	THS	THS	NSCP, RW, S
12	18.00	MB	Threadlock	6" Straight	Garnix	THS	THS	NSCP, RW, S
13	18.00	MB	Threadlock	7½" Straight	Garnix	THS	THS	NSCP, RW, S
312	18.00	MB	Threadlock	6" Straight	Garnix	THS	THS	NSCP, RW, S
33	15.00	DT	Threadlock	6" Straight	Garnix	THS	THS	NSCP, RW, S
34	15.00	DT	Threadlock	7½" Straight	Garnix	THS	THS	NSCP, RW, S
25	18.00	MB	Threadlock	6" Straight	Garnix	THS	THS	NSCP, RW, S
29	7.50	LB	NSCP, C&R	6" Straight	Garnix	THS	THS	NSCP, RW, S
36	15.00	DT	Screwlock, BNS	6" Half Wells	Garnix	THS	THS	BNS, RW, S
37	15.00	DT	Screwlock, BNS	6" Half Wells	Garnix	THS	THS	BNS, RW, S
336	15.00	DT	Screwlock, BNS	6" Half Wells	Garnix	THS	THS	BNS, RW, S
23	12.00	LB	Screwlock, CP	6" Rev. HW	Garnix	THS	THS	NSCP, RW, S
24	12.00	LB	Screwlock, CP	7½" Rev. HW	Garnix	THS	THS	NSCP, RW, S
323	12.00	LB	Screwlock, CP	6" Rev. HW	Garnix	THS	THS	NSCP, RW, S
46	10.00	DT	Screwlock, CP	6" Half Wells	Garnix	THSCP	THSCP	NSCP, RW
47	10.00	DT	Screwlock, CP	6" Rev. HW	SS	SS	SS	NPB, RW
77	7.50	LB	Screwlock, CP	5" Rev. HW	Garnix	THS	THS	CPB
76	6.00	LB	CPB, C&R	5" Full Wells	THS	THS	THS	CPB

Rod #	$	Cane	Reelseat	Grip	Stripper	Snake	TT	Ferrule
Best–O–Luck Series:								
59	5.00	DT	CPB, C&R	5" Full Wells	THS	THS	THS	CPB
58	4.50	LB	CPB, C&R	5" Full Wells	THS	THS	THS	CPB
57	3.50	N	CPB, C&R	5" Half Wells	THS	THS	THS	CPB
56	3.15	N	NPB, C&R	5" Half Wells	THS	THS	THS	NPB
55	2.50	N	NPB, C&R	5" Half Wells	THS	THS	THS	NPB

South Bend Bait Co. Wrap Specs 1936

Rod #	Main Wraps	Tipping	Intermediates	Notes	Signature
30/31	Yellow #5	Red #10	Black #1	Standard intermediates	Yes
119/120	Green #24	Orange #8	Green #24	Wide spacing	Yes
27/28	B&W Jasper	Orange #8	N/A		Yes
12/13/312	Red #10	Black #1	N/A		Yes
33/34	Purple #16	Orange #8	N/A		Yes
25	Orange #8	Black #1	Black #1	Wide spacing	No
29	Yellow #5	Green #24	N/A		No
36/37/336	Brown #28	Primrose #15	N/A		No
23/24/324	B&O Jasper	N/A	N/A		No
46/47	Primrose #15	Orange #8	N/A		No
77	Yellow #5	Green #24	N/A		No
76	Orange #8	Purple #16	N/A		No
59	Yellow #5	Blue #17	N/A		No
58	Red #10	Green #24	N/A		No
57	Green #24	N/A	N/A		No
55/56	Red #10	N/A	N/A		No

South Bend Bait Co. Hardware Specs 1942

Rod	Action	$	Cane	Reelseat	Grip	Stripper	Snake	TT	Ferrule
12	T	22.00	MB	LL-green/chrome	Comficient	SS	THS	PC	NSCP, RW, S
13	B	22.00	MB	LL-green/chrome	Comficient	SS	THS	PC	NSCP, RW, S
312	DF	22.00	MB	LL-green/chrome	Comficient	SS	THS	PC	NSCP, RW, S
23	T	16.50	MB	LL-tan/chrome	Comficient	SS	THS	PC	NSCP, RW, S
24	B	16.50	MB	LL-tan/chrome	Comficient	SS	THS	PC	NSCP, RW, S
323	DF	16.50	MB	LL-tan/chrome	Comficient	SS	THS	PC	NSCP, RW, S
46	T	15.00	DT	LL-maroon/chrome	Comficient	SS	THS	THS	NSCP, RW, S
47	B	15.00	DT	LL-maroon/chrome	Comficient	SS	THS	THS	NSCP, RW, S
346	DF	15.00	DT	LL-maroon/chrome	Comficient	SS	THS	THS	NSCP, RW, S
177	T	12.50	MB	LL-black/chrome	Comficient	SS	THS	THS	NSCP, RW, S
77	B	12.50	MB	LL-black/chrome	Comficient	SS	THS	THS	NSCP, RW, S
377	DF	12.50	MB	LL-black/chrome	Comficient	SS	THS	THS	NSCP, RW, S
159	T	9.35	DT	LL-yellow/chrome	Comficient	Garnix	THS	THS	NS, RW, S
59	B	9.35	DT	LL-yellow/chrome	Comficient	Garnix	THS	THS	NS, RW, S
359	DF	9.35	DT	LL-yellow/chrome	Comficient	Garnix	THS	THS	NS, RW, S
29	B	11.00	MB	LL-black/chrome	Comficient	SS	THS	THS	NSCP, RW, S
290	DF	11.00	MB	LL-black/chrome	Comficient	SS	THS	THS	NSCP, RW, S
58	B	7.75	MB	SDL, CPB	7½" Half Wells	Garnix	THS	THS	NSCP, RW, S
358	DF	7.75	MB	SDL, CPB	6½" Half Wells	Garnix	HS	THS	NSCP, RW, S
57	B	7.00	DT	SDL, CPB	7½" Half Wells	Garnis	HS	HS	CPB, RW
357	DF	7.00	DT	SDL, CPB	6½" Half Wells	THS	PC	HS	CPB, RW
55/56	B	6.50	N	CPB, C&R	6½" Half Wells	THS	PC	THS	CPB, RW

South Bend Bait Co. Hardware Specs 1942

Rod	Action	$	Cane	Reelseat	Grip	Stripper	Snake	TT	Ferrule
51	B	22.00	N	LL-white/black	7½" Rev. HW	CPSS	THS	P	BNS, HW, S, WP
50	B	15.00	MB	LL-maroon/chrome	7½" Rev. HW	Garnix	THS	P	NSCP, HW, S, WP
424	S	19.25	N	SDL, CPB	2-Hand	CP	HS	P	NSCP, HW, S, WP
26	DF	6.50	LB	CPB, C&R	6" Cigar	Garnix	THS	P	NSCP, RW, S, WP
260	T	6.50	LB	CPB, C&R	6" Cigar	Garnix	THS	P	NSCP, RW, S, WP

Cross DoubleBuilt Rods by South Bend

Rod	$	Cane	Reelseat	Grip	Stripper	Snake	TT	Ferrule
30/330	38.50	B/N	LL-white/black	6" Mild Wells	CPSS	THS	P	NSCP, HW, S, WP
31	38.50	B/N	LL-white/black	7½" Mild Well	CPSS	THS	P	NSCP, HW, S, WP
119/319	33.00	N	LL-yellow/black	6" Rev. HW	SS	THS	P	NSCP, HW, S, WP
120	33.00	N	LL-yellow/black	7½" Rev. HW	BSS	THS	P	NSCP, HW, S, WP
27/327	27.50	MB	LL-black/black	6" Rev. HW	BSS	THS	P	NSCP, HW, S, WP
28	27.50	MB	LL-black/black	6" Rev. HW	BSS	THS	P	NSCP, HW, S, WP
36/336	20.00	DT	LL-black/black	6" Half Wells	BSS	BTHS	PB	BNS, HW, S, WP
37	20.00	DT	LL-black/black	7½" Half Wells	BSS	BTHS	PB	BNS, HW, S, WP
25	16.50	MB	LL-green/black	6" Mild Cigar	SS	THS	P	BNS, HW, S, WP

South Bend Bait Co. — Wrap Specs — 1942

Models 12, 13, 312
B&G Jasper #XXXXX
Tipped Yellow #5

Models 23, 24, 323
Brown #28
Tipped White #2

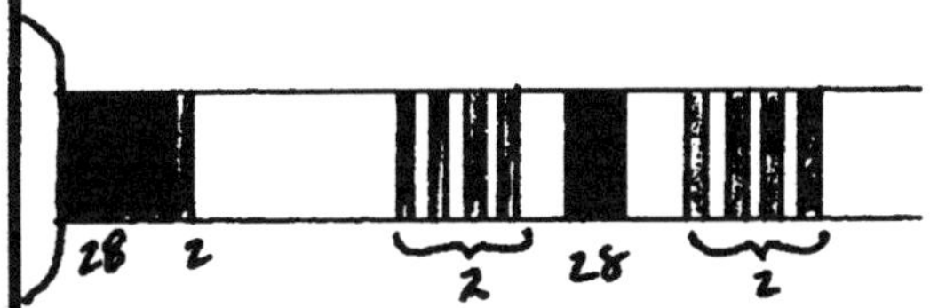

Models 46, 47, 346
Red #10
Tipped White #2

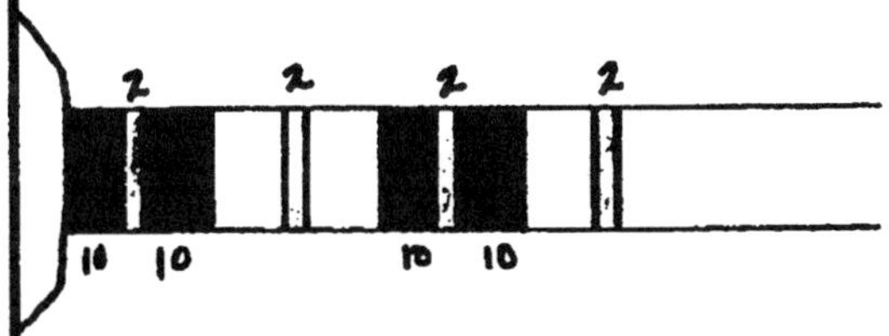

Models 177, 77, 377
Green #24
Tipped Yellow #5

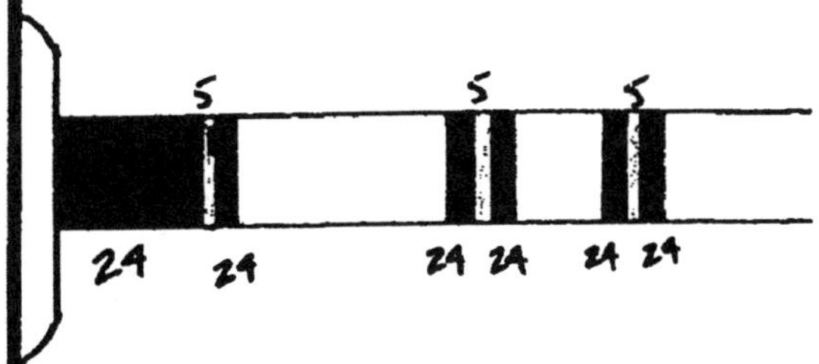

Models 159, 59, 359
B&W Jasper #X
Plain - No Tipping

Models 29, 290
B&O Jasper #XX
Plain - No Tipping

South Bend Bait Co. — Wrap Specs — 1942

Models 58, 358
White #2
Tipped Blue #17 at each end

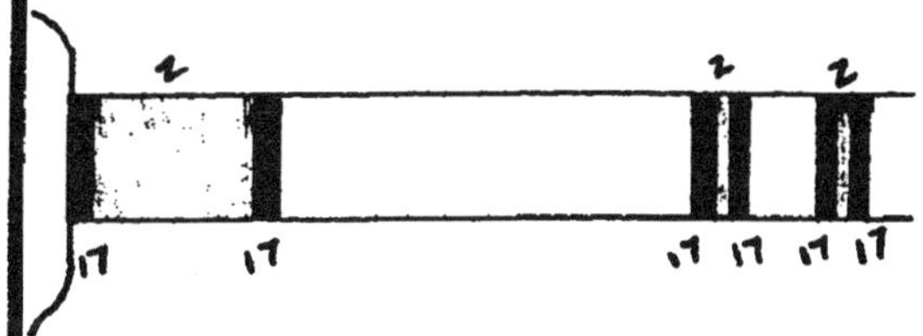

Models 57, 357
Rainbow #Z

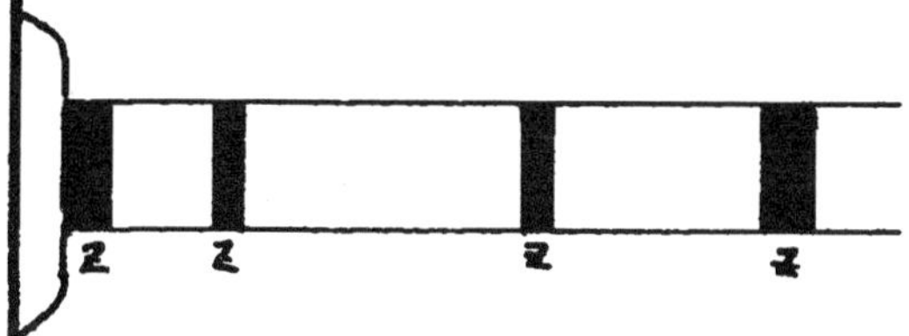

Models 55, 56
Green #23

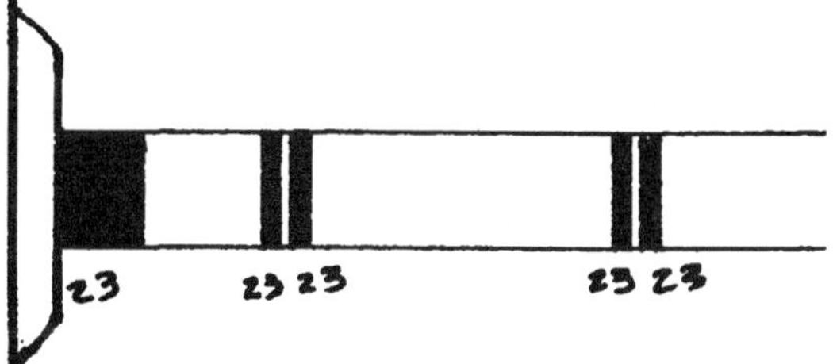

Model 51
Black #1
Tipped White #2

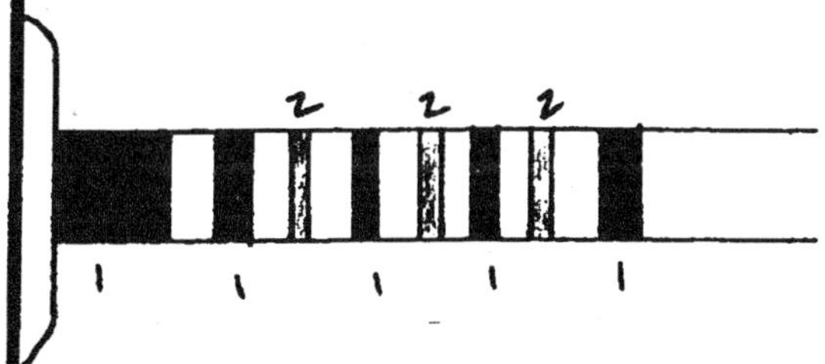

Model 50
Red #10
Tipped White #2

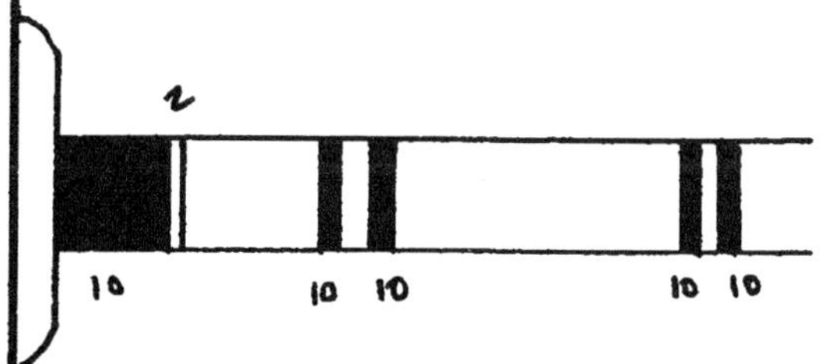

Models 26, 260
Rainbow #Z

Cross DoubleBuilt by South Bend Wrap Specs 1942

Models 30, 31, 330
Black #1
Tipped White #2

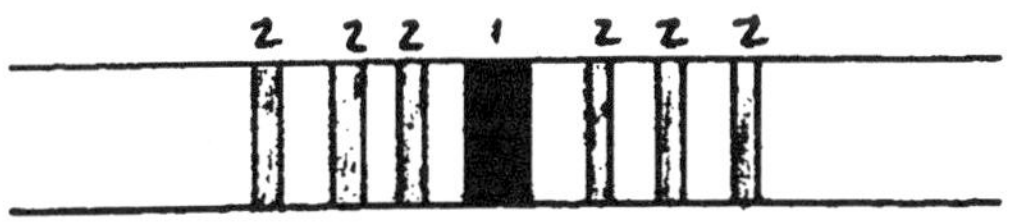

3" Above Grip

Models 119, 120, 319
Yellow #5
Tipped Black #1

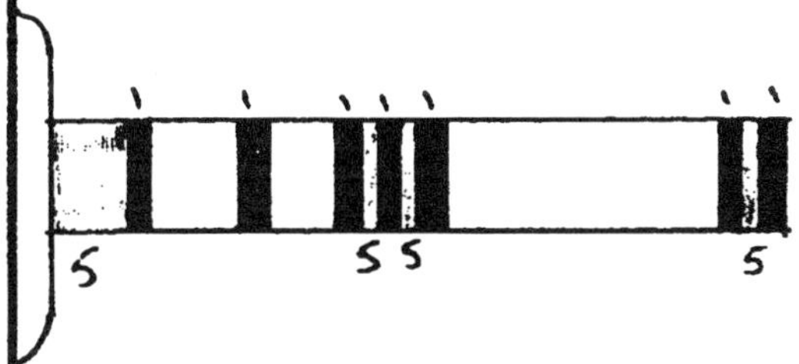

Models 27, 28, 327
B&W Jasper #X
Tipped Orange #8

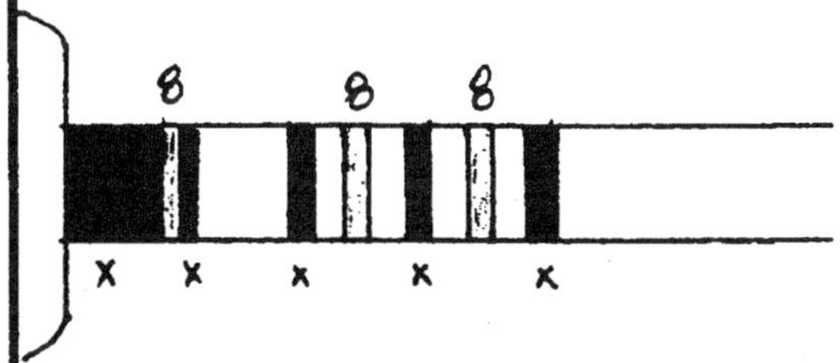

Evenly Spaced

Models 36, 37, 336
B&W Jasper #X
Tipped Black #1

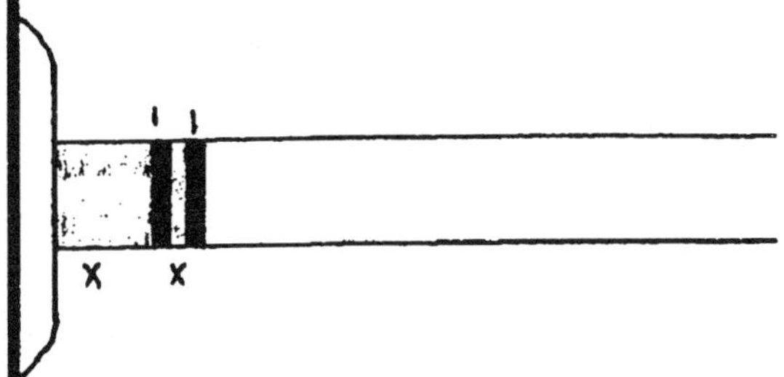

Model 25
Orange #8
Tipped Black #1

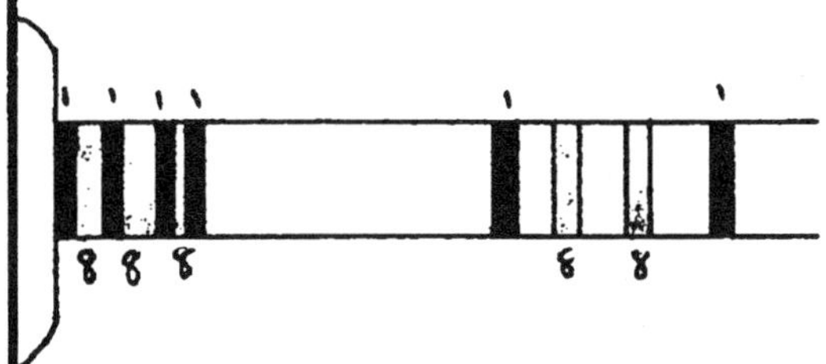

South Bend Bait Co. — Hardware Specs — 1949

Rod #	$	Cane	Reelseat	Grip	Stripper	Snake	TT	Ferrule
53	35.00	N	Litelock	Comficient	SS	THS	PC	NSSC, RW, S
24	25.00	N	Litelock	Comficient	SS	THS	PC	NSSC, RW, S
323	25.00	N	Litelock	Comficient	SS	THS	PC	NSSC, RW, S
47	20.00	F	Litelock	7½" Comficient	SS	THS	PC	NSSC, RW, S
346	20.00	F	Litelock	Comficient	SS	THS	PC	NSSC, RW, S
59	15.00	F	Litelock	Comficient	Garnix	TS	TS	NSSC, RW, S
359	15.00	F	Litelock	Comficient	Garnix	TS	TS	NSSC, RW, S
290 2/2	16.00	N	Litelock	Comficient	SS	TS	TS	NSSC, RW, S
496 2/1	20.00	N	Litelock	10" Straight, spin	Spin	CP, spin	SSCP	NSSC, RW, S

South Bend Bait Co. Hardware Specs 1951

Rod #		$	Cane	Reelseat	Grip	Stripper	Snake	TT	Ferrule
24		27.95	MB	PL - Cork	7' Mild HW	SS	SS	PC	NSCP, RW, S
323		27.95	MB	PL - Cork	7" Mild HW	SS	SS	PC	NSCP, RW, S
47		22.95	F	PL - Maroon	7" Conficient	SS	SS	PC	NS, RW, S
346		22.95	F	PL - Brown	7" Comficient	SS	SS	PC	NS, RW, S
59		16.95	F	PL - Brown	6" Comficient	CP	SS	CP	NS, RW, S
359		16.95	F	LL - Black	6" Comficient	CP	SS	CP	NS, RW, S
51	2/2	25.00	MB	Screwlock, PC	7" Comficient	CP	SS	CP	NSCP, RW, S
57		11.95	F	Screwlock, PC	6" Composition	THS	THS	THS	NS, RW
357		11.95	F	Screwlock, PC	6" Composition	THS	THS	THS	NS, RW
55		8.95	F	Screwlock, PC	6" Composition	THS	THS	THS	NS, RW
355		8.95	F	Screwlock, PC	6" Composition	THS	THS	THS	NS, RW
290	2/2	16.95	MB	PL - Brown	6" Comficient	SS	THS	THS	NS, RW, S
260	2/1	8.95	MB	Screwlock, NP	6" Composition	THS	THS	THS	NS, RW
291	4/1	16.95	MB	PL - Brown	6½" Comficient	SS	THS	THS	NS, RW, S
469	2/1	18.95	F	LL - twin rings	10" Spin style	CP, spin	CP, spin	SSCP	NS, CP, RW, S

South Bend Bait Co. **Wrap Specs** **1951**

Rod #	Thread	Main Wraps	Tipping	Notes
24/323	Nylon	Black #1	N/A	Hookkeeper
47/346	Nylon	Maroon #13	Yellow #5	Hookkeeper
59/359	Silk	B&W Jasper #X	N/A	Hookkeeper
57/357	Nylon	Red #10	N/A	Hookkeeper
55/355	Nylon	Brown #28	N/A	
51 2/2	Nylon	Red #10	Black	Hookkeeper
290 2/2	Silk	B&O Jasper #XX	N/A	Hookkeeper
260 2/1	Nylon	Red #10	N/A	
291 4/1	Silk	Olive #21	Red #10	Hookkeeper

South Bend Reel Seat Variations

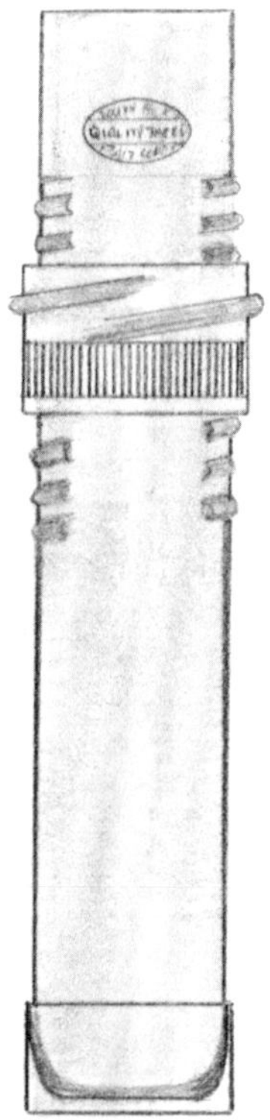

SCREW-LOCK

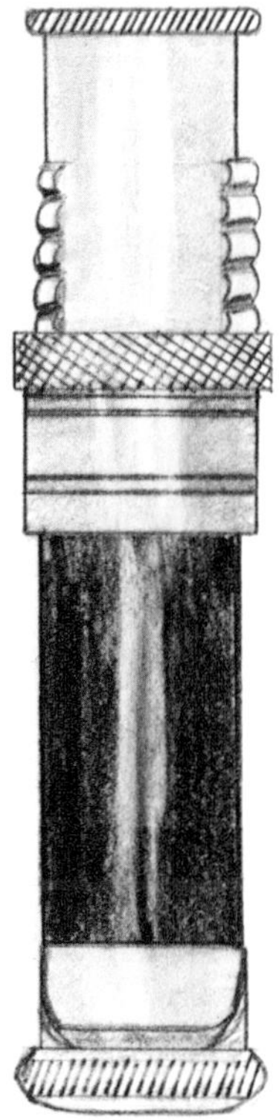

THREAD-LOCK

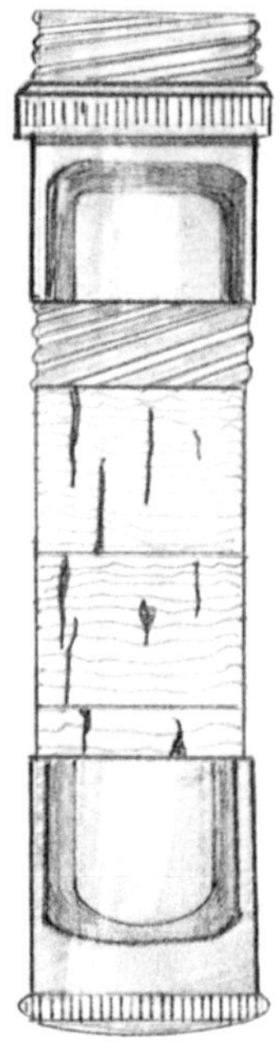

POSITIVE LOCK

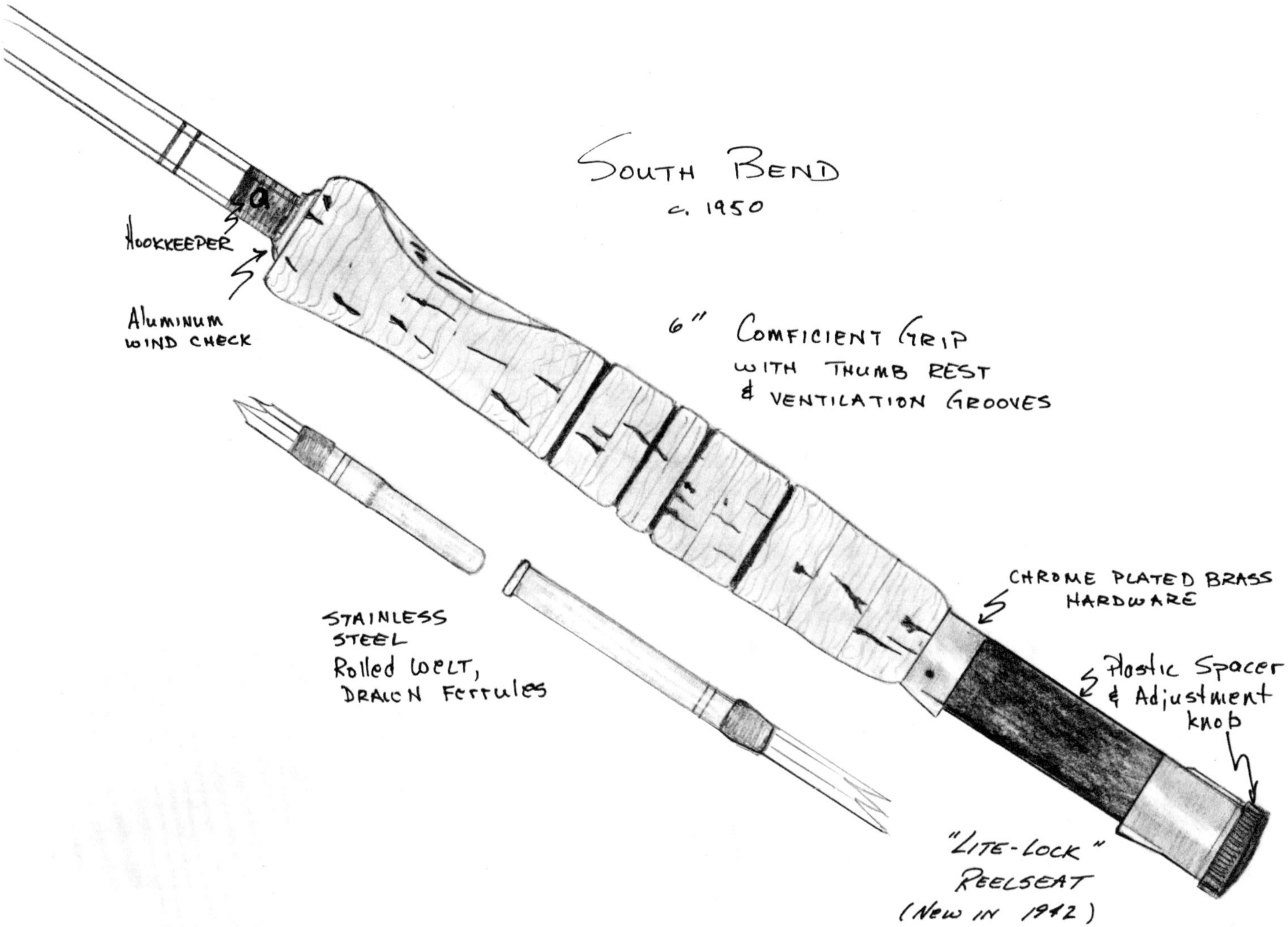
South Bend
c. 1950
Hookkeeper
Aluminum
wind check
6" Comficient Grip
with thumb rest
& ventilation grooves
Stainless
steel
Rolled welt,
drawn ferrules
Chrome plated brass
hardware
Plastic Spacer
& Adjustment
knob
"Lite-Lock"
Reelseat
(New in 1942)

Wright & McGill Rod Company

After World War II, Wright & McGill made an arrangement with Goodwin Granger's family to lease the Goodwin Granger Co., and to eventually buy it. Wright & McGill had been in the tackle business since the 1920's, but had never produced rods. The Wright & McGill Rod Company was formed in 1946, and rehired many former Granger emmployees as well as many new employees to meet the post-war demand. The first year, there were no changes in the rods themselves. The reelseat marking was changed to #5 immediately, as there were no left-over reelseats from the pre-war production. The Wright & McGill name was rubber stamped on the #5 flat in a style identical to that of the Granger grade designation which continued to appear on flat #1.

In 1947, in the second year, W&M began to make changes. The Champion model was dropped and replaced by the Stream and Lake. This model had red *nylon* wraps with no tipping, and the Granger uplocking reelseat. Other changes by W&M were simply cosmetic, but can be used to help tell the earlier rods from the later ones. The Victory model was the first to see a cosmetic change. Initially the border wraps were dropped leaving black/orange Jasper tipped black. This was in 1947. The next year, all of the black trim was eliminated, and the wraps were simple black/orange Jasper. Also in 1947, the DeLuxe grade lost one of its border wraps, leaving only two borders (plus the tipping) on each wrap.

In 1951, more changes took place to simplify the wraps. The DeLuxe lost all border wraps, leaving its wraps black/white Jasper tipped gold. The Favorite's signature wrap was cut down to three gold bands in the signature (down from six).

At some point, both of the rodshaft markings were changed. Though it is not certain, it seems that this probably took place in 1951. The change was twofold; the direction of the lettering was reversed so that it read toward the grip, rather than away from the grip. Also, the rubber stamp was changed to a heavier and larger script which made the marking much easier to read. The Water Seal rods were added to the line in 1952. This was W&M's impregnated series of rods. They were produced in two grades, FA and FB. The FA was the higher grade, and had ferrules and a Granger patent reelseat made from a new alloy called *Monel*. It was supposed to be superior to NS, but only turned out to be more expensive. The FA rods were wrapped in burgundy nylon, tipped gold and were packed in a burgundy satin bag and burgundy anodized tube with gold anodized cap. The FB had NS hardware and was wrapped in gold tipped burgundy, packed in gold bag and gold anodized tube with burgundy cap. A spinning rod model of the Water Seal was also available. All models had a rectangular decal that denoted the rod to be a "Wright & McGill / Water Seal Hand-Made Super Rod". The Water Seal rods were available as late as 1960.

Wright & McGill Co.

FISHING TACKLE MANUFACTURERS
1463 York Street -:- Denver 6, Colorado
U.S.A.

#5

MADE BY
WRIGHT & MCGILL ROD CO.
DENVER, COLO., U.S.A.

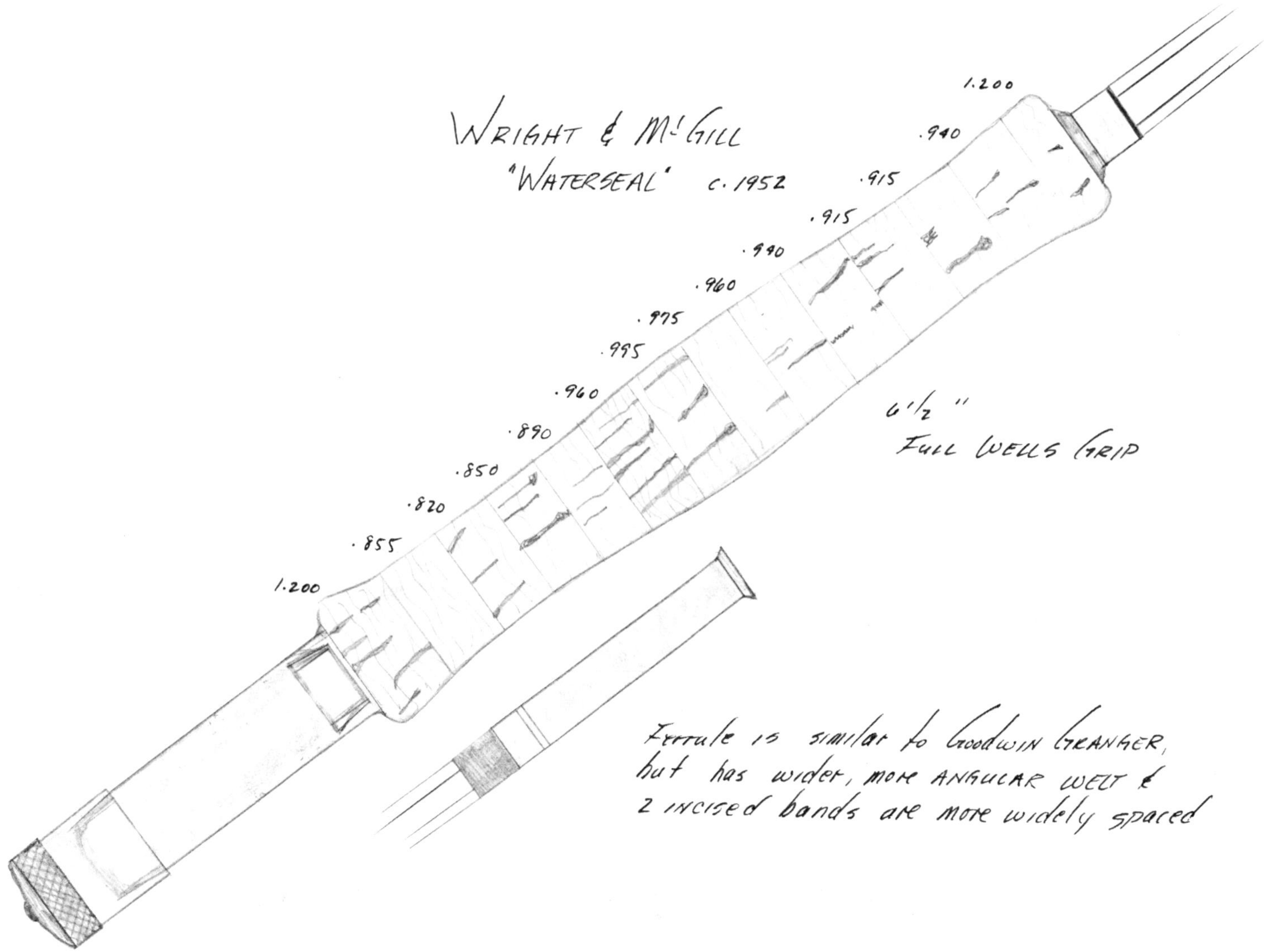
WRIGHT & McGILL
"WATERSEAL" c. 1952
1.200
.940
.915
.915
.940
.960
.975
.995
.960
.890
.850
.820
.855
1.200
6 1/2"
FULL WELLS GRIP
Ferrule is similar to Goodwin Granger, but has wider, more angular welt & 2 incised bands are more widely spaced

Wright & McGill Trade Rods

When Wright & McGill took over the Granger operation after WW II they inherited many established dealers of Granger rods. Since Granger dealers were often promised an exclusive marketing right for an area, this meant that W&M could sell Granger rods to only a few dealers in any area. To get around this, W&M created two trade names for selling their rods. These trade rods by Wright & McGill were very similar to the regular Granger rods. In fact, the rods were the same rods made to be sold as Wright & McGill Granger rods. They carried the same ferrules, and usually only the wraps and reelseats were changed.

The most common of the W&M trade rods were those sold under the Old Faithful Rod Company name. The Old Faithful name appeared only on the rod tube label, and only the model name was rubber stamped on the rodshaft. Most of these rods were of the same quality level as the W&M Stream & Lake or the Victory rods. The lowest of the grades was called the Challenger. This rod used a screw, downlocking reelseat of maroon anodized aluminum. The grip was of the same style and size of the regular rods, and was usually made of the same cork rings intended for the lower grade W&M Grangers, but was sometimes made of the cheaper cross-cut cork rings. Guide wraps for these rods were chocolate brown – the same color thread used as tipping for a Granger Aristocrat (Brown #28 on the thread chart). These wraps were plain, with no tipping or ornamental wraps at all.

Another model for Old Faithful Rod Co. was called the Thoroughbred, and was of the same grade as the Granger Victory. This rod was identical to the Challenger, but was wrapped with the tan color used as the main wraps on the Granger Aristocrat, which is Beige #27 on the thread chart. The last of the Old Faithful rods was called the True Action, and is like the others, but of indiscriminate quality. The wraps for this model are usually one of the above mentioned colors.

The rarest of the W&M trade rods were labeled Aladdin Rod Co. The Aladdin name is rubber stamped on the rodshaft, as is the model name *Tried & True*. This model rod was the exact equivalent of the Granger Special. It used the same *Granger green* wraps, and had the Granger patented internal uplocking reelseat. The reelseat may or may not have the stamped markings for the Wright & McGill Rod Co. These rods were not *seconds*, as were some of the Old Faithful rods. They were sold mostly by Whitney Sporting Goods in Denver. Tried & True was a trademark of the old Tritch Hardware Company which had been absorbed by Whitney Sporting Goods in the 1930's.

The only other common Wright & McGill trade rods were those made for Marshall Field Company of Chicago. These were called the One-Striper, Two-Striper and Three-Striper. They were equivalent to the Granger Victory, Special and Aristocrat, and the wraps were the same as for the equivalent Granger rods. These rods were marked by stripes of *Granger green* thread just above the winding check wrap; the number of stripes indicating the grade.

Private Label Bamboo Rods

Private Label Rod Decals

NEW
KIFFE
YORK.

A F

S.D. & G.
NEW YORK

TRADE MARK

ABERCROMBIE & FITCH CO.
57 READE STREET NEW YORK

Schoverling, Daly & Gales

NEW YORK SPORTING GOODS CO

N. Y. S. G. CO

N. Y. Sporting Goods Co.

V. L. & D.

VON LENGERKE & DETMOLD INC

William Mills & Son

VON LENGERKE & ANTOINE
"V. L. & A."
CHICAGO

OSPREY
HIGHEST GRADE
LOU. J. EPPINGER CO., DETROIT, MICH.

THE Whitney
SPORTING GOODS CO.
2535 WALNUT ST., DENVER, COLO

THE
SIGN OF THE
LEAPING DOLPHIN
"Fishing Tackle that's Fit for Fishing"

Courtney Riley Cooper

Courtney Ryley Cooper was a Colorado outdoors writer who gained national prominence. He was a regular contributor to many magazines and authored several books. Cooper's most popular era was from c. 1926 to about the time of WW II. During this period, his name was associated with several tackle companies. The first connection was with Winchester's Fishing Rod division. Cooper was the celebrity spokesman for Winchester rods from about 1926 to about 1931. His name was used by Winchester in advertising, but did not appear as a model name for the rods.

In about 1931, upon the sale of Winchester's fishing rod division to Horrocks-Ibbotson, Cooper switched his alliance to South Bend. At South Bend, Cooper's name was seen on at least two Cross Double-built models. He was also seen in some advertising for South Bend.

Being a resident of Colorado, Cooper also promoted many Colorado tackle products. He promoted Wright & McGills's *Wiltless Wing* Flies and wrote for the Dave Cook Sporting Goods catalog. Dave Cook is one of Denver's oldest sporting goods companies, and in the late 1930's produced a very large and thorough catalog that contained informative articles as well as a wide variety of sportsmen's toys. After WW II, Denver rodmaker Fred DeBell began to market rods that carried the name : Courtney *Riley* Cooper. Note the different spelling of the middle name. These rods were assembled from parts DeBell bought from Wright & McGill Rod Company and the Phillipson Rod & Tackle Company of Denver. They were sold through Dave Cooks, who may have still had rights to the use of the name. The different spelling makes it easy to tell those made by DeBell from those sold earlier by South Bend.

Lyon & Coulson

Lyon & Coulson of Buffalo, N. Y. sold rods made by other manufacturers. L&C never made their own rods. It is hard to identify the maker of a few L&C rods, but most are Heddon products and with the exception of some wraps they are identical to the rods sold by Heddon. From the early 1930's until the late 1940's, Heddon was L&C's rod supplier. In 1937, these are the L&C grade names, selling prices and Heddon equivelents:

DeLuxe - $50.00 - Heddon #35 Peerless
Regent - $28.00 - Heddon #20 Bill Stanley's Favorite
Imperial - $22.50 - Heddon #17 Black Beauty
Crown - $16.50 - Heddon #14 Thorobred

Within the grades were model numbers that indicated both the length of the rod and the action:

DELUXE:
No. 00 7½', 2/2 Trout Action
No. 010 Special Trout Action
No. 01 8', 2/2 Trout Action
No. 011 Special Trout Action
No. 1 8', 3/2 Trout Action
No. 11 Special Trout Action
No. 2 8½', 3/2 Trout Action
No. 12 Special Trout Action
No. 3 9', 3/2 Trout Action
No. 13 Special Trout Action
No. 23 9', 3/2 Bass Action
No. 34 9½' Salmon Rod
No. 24 9½', 3/2 Bass Action
No. 35 10' Salmon Rod

REGENT
No. 0100 7½', 2/2 Trout Action
No. 0110 Special Trout Action
No. 0101 8' , 2/2 Trout Action
No. 0111 Special Trout Action
No. 0102 8' , 3/2 Trout Action
No. 0112 Special Trout Action
No. 0103 8½', 3/2 Trout Action
No. 0113 Special Trout Action
No. 123 9', 3/2 Bass Action
No. 134 9½' Salmon Rod
No. 124 9½', 3/2 Bass Action
No. 135 10' Salmon Rod

IMPERIAL
No. 0200 7½', 2/2 Trout Action
No. 223 9', 3/2 Bass Action
No. 201 8', 3/2 Trout Action
No. 224 9½', 3/2 Bass Action
No. 202 8½'. 3/2 Trout Action
No. 203 9', 3/2 Trout Action

CROWN
No. 0300 7½', 2/2 Trout Action
No. 323 9', 3/2 Bass Action
No. 301 8', 3/2 Trout Action
No. 324 9½', 3/2 Bass Action
No. 302 8½'. 3/2 Troout Action
No. 303 9', 3/2 Trout Action

The grade name and model number were inked on the rodshaft.

Edward K. Tryon Co. Rods

Jay Harvey Rods

"Jay Harvey" was a trade-name used by the Edward K. Tryon Co. of Philadelphia to compete with names such as Bill Stanley (Heddon), Marvin Hedge (Heddon), Jim Haywood (Whitney Sporting Goods, Denver), Courtney Riley Cooper (South Bend), Cross Rods (South Bend). The name was first used in the early 1930's, and continued to be used until about 1950. It is not clear whether there really was someone named Jay Harvey, or if it was a a name invented by Tryon's marketing department.

One story says that J. Harvey Tryon was an early principal in this old firm, which goes back to 1811. None of the casting tournament records from the 1920's or 1930's show this name as a winner or even as a participant. (Many rods were marketed as being used by famous tournament casters, including names like Marvin Hedge and Bill Stanley.) The earliest Jay Harvey marked rods were made by Montague. There was one grade only, which featured all NS hardware, a sliding band reelseat, Wells grip, swelled butt and honey colored cane. These rods are usually seen with Spring Green wraps. The wraps are very plain and the rod has closely spaced full intermediates. The name Jay Harvey is stippled into the NS reelseat.

Beginning about 1934, Jay Harvey rods were made by Heddon. The reelseat, ferrules, and browntone Tempered Cane of Heddon rods are the same used on Jay Harvey Rods. The grip shape is that of the Heddon #10 - a straight grip with a slight taper at the cane end. This is why so many authorities have believed that Jay Harvey was a Heddon rodmaker. Heddon records indicate that there was never an employee named Jay Harvey, and that the name had no other connection to Heddon.

Jay Harvey rods by Heddon were made in five grades and seventeen models. All grades were named for former U.S. Presidents. In order from highest to lowest grade listed with the equivelent in the Heddon line they are:

Washington (Heddon #50)
Jefferson (Heddon #35)
Jackson (Heddon #20)
Cleveland (Heddon #13)
Lincoln (Heddon #10)

The grade name and the model number were inked on the rodshaft. Each model ended in a letter that designated the type of action:

R = Regular Trout
L = Light Trout
B = Bass Fly
D = Dry Fly

All Jay Harvey rods were three-piece rods with one exception, the Lincoln X11R which was a 2-piece, 7½'.

Wraps for the Pre WW II Heddon Jay Harvey rods were listed only as *Tournament Style* which means that there were no ornamental wraps, only those which were necessary at guides and ferrules. Traditional colors for Tournament rods were Black & White Jasper, Gold, or Red. A later Jay Harvey model was called the TriColor. Post WW II Jay Harvey rods had new standardized wrap colors. These use the same colors as Heddon silks:

Jackson = Green #23 tipped Gold #7, ferrules Japanned dark Green
Cleveland = Gold #7 tipped Brown #30
Lincoln = Red #10

Examples of the Washington and Jefferson grade Jay Harvey rods have not been available for examination.

EDW. K. TRYON COMPANY

Sporting Goods and Hardware 817-819 Arch Street, Philadelphia, Pa.

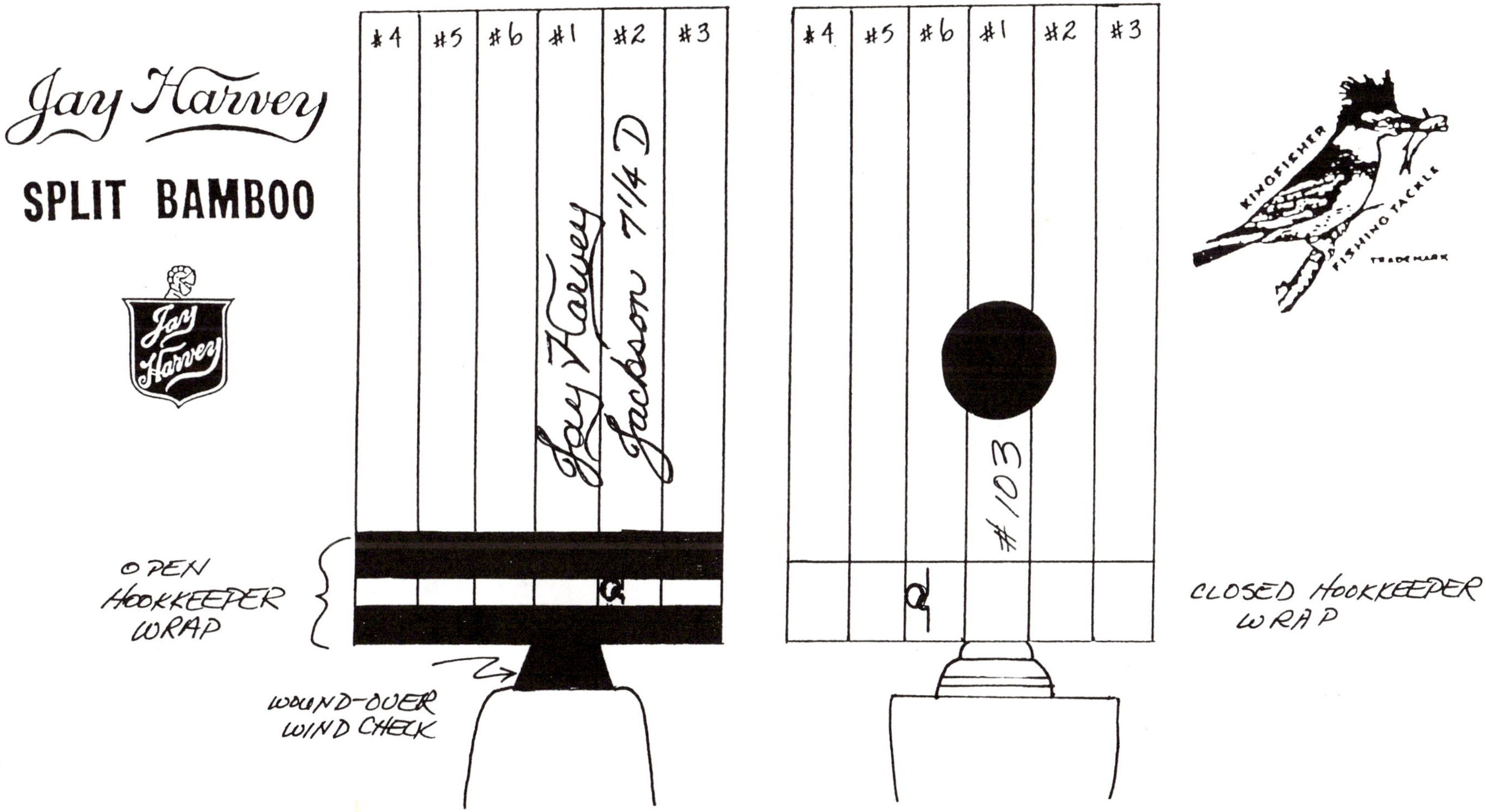

Jay Harvey Rods　Hardware Specs　1938

Model	#	$	Reelseat	Grip	Stripper	Snake	TT	Ferrules
Washington	15R	50.00	SDL, Walnut	7" straight	A	HS	A/HS	Heddon NS, GM, HW
Washington	15½D	50.00	SDL, Walnut	7" straight	A	HS	A/HS	Heddon NS, GM, HW
Jefferson	13R	35.00	SDL, Walnut	7" straight	A	HS	A/HS	Heddon NS, GM, HW
Jefferson	13½D	35.00	SDL, Walnut	7" straight	A	HS	A/HS	Heddon NS, GM, HW
Jackson	7R	25.00	UL, Cedar	7" straight	A	HS	A/HS	Heddon NS, HW
Jackson	7¼D	25.00	UL, Cedar	7" straight	A	HS	A/HS	Heddon NS, HW
Jackson	7½B	25.00	UL, Cedar	7" straight	A	HS	A/HS	Heddon NS, HW
Jackson	7¾ L	25.00	UL, Cedar	7" straight	A	HS	A/HS	Heddon NS, HW
Cleveland	X3R	18.00	UL, Pyralin	7" straight	A	HS	A/HS	Heddon NS, RW
Cleveland	X3¼D	18.00	UL, Pyralin	7" straight	A	HS	A/HS	Heddon NS, RW
Cleveland	X3½B	18.00	UL, Pyralin	7" straight	A	HS	A/HS	Heddon NS, RW
Cleveland	X3¾L	18.00	UL, Pyralin	7" straight	A	HS	A/HS	Heddon NS, RW
Lincoln	X1R	15.00	UL, Tenite	7" straight	HS	HS	HS	Heddon NS, RW
Lincoln	X1¼D	15.00	UL, Tenite	7" straight	HS	HS	HS	Heddon NS, RW
Lincoln	X1½D	15.00	UL, Tenite	7" straight	HS	HS	HS	Heddon NS, RW
Lincoln	X1¾D	15.00	UL, Tenite	7" straight	HS	HS	HS	Heddon NS, RW
Lincoln	X11R	15.00	UL, Tenite	7" straight	HS	HS	HS	Heddon NS, RW

Kingfisher Rods

Kingfisher Rods were sold exclusively by the E. K. Tryon Co. of Philadelphia. Since the Jay Harvey rods were the high grade label for E. K. Tryon, the Kingfisher rods were of middle to very low grades. These were typical *Private Label* trade rods made by Montague.

Kingfisher rods had the usual characteristics of Montague rods of the era, and usually the only difference was the Kingfisher decal in place of the Montague decal. The Kingfisher decal is usually a circle about 7/8" in diameter. At the center of the decal is a Kingfisher holding a fish in its beak. The words "Trade / Mark" flank the image. The Kingfisher is slate blue with white highlights. All hardware for Kingfisher rods is nickel-plated brass. Most of the ferrules are typical Montague shouldered, rolled welt ferrules, though some of the lowest grades had no welt. Reelseats were of two types, either a sliding band on lower grades or a screw, down-locking seat on the middle to higher grades. See the drawings of Montague reelseats for more information.

The color of the cane varies with the model, as it does for all rods by Montague. See the chart for details. Wrap colors were not specified. Most of the rods had very simple wraps, and if there were intermediates, they were ornamental and were spaced widely. If two colors are used, the color of the intermediates usually alternates. Use typical Montague colors.

After WW II, Kingfisher Rods were made by Heddon. With Kingfisher rods, the higher the model #, the higher the grade. There were fewer models made, and the lowest grade was the #103. When this rod was introduced, it had no equivelent model in the Heddon rod line, but when it was added to Heddon's catalog, it was the #8. The Heddon Kingfisher rods usually have no decal, but have "Kingfisher #103" inked on the #1 flat. They are easy to spot as lower grade Heddons as they look like their Heddon counterparts; most have straight grips, exposed winding check, and Heddon's down-locking tenite reelseat. All fly rods made by Heddon have browntoned Tempered Cane. Ferrules will be straight, NS, and have rolled ferrules.

Kingfisher Rods — Hardware Specs — 1938

Rod #	$	Action	Reelseat	Cane	Grip	Stripper	Snake	TT	Ferrules
127	25.00	RT	NPB, SDL	Browntone	Slender HW	HS	HS	HS	SH, NPB, RW
108	6.50	RT	NPB, SDL	Browntone	Fat Half Wells	G	HS	G/HS	SH, NPB, RW
115	12.00	RT	NPB, SDL	Flamed	Fat Half Wells	G	HS	G/HS	SH, NPB, RW
115½	12.00	LT	NPB, SDL	Flamed	Fat Half Wells	G	HS	G/HS	SH, NPB, RW
113	10.00	RT	NPB, SDL	Flamed	Half Wells	G	HS	G/HS	SH, NPB, RW
113½	10.00	LT	NPB, SDL	Flamed	Half Wells	G	HS	G/HS	SH, NPB, RW
0113½	10.00	LT	NPB, SDL	Flamed	Half Wells	G	HS	G/HS	NPB, RW, S
111	9.00	RT	NPB, SDL	Browntone	Half Wells	G	HS	G/HS	SH, NPB, RW
111½	5.50	LT	NPB, SDL	Browntone	Half Wells	G	HS	G/HS	SH, NPB, RW
109	7.50	RT	NPB, SDL	Browntone	Straight	G	HS	G/HS	SH, NPB, RW
109½	7.50	LT	NPB, SDL	Browntone	Straight	G	HS	G/HS	SH, NPB, RW
107	5.00	RT	NPB, SDL	Browntone	Straight	HS	HS	G/HS	SH, NPB, RW
107½	5.00	LT	NPB, SDL	Browntone	Straight	HS	HS	G/HS	SH, NPB, RW
105	4.00	RT	NPB, SDL	Browntone	Cigar	HS	HS	HS	SH, NPB, RW
105½	4.00	LT	NPB, SDL	Browntone	Cigar	HS	HS	HS	SH, NPB, RW
103	3.50	RT	NPB, SDL	Browntone	Cigar	HS	HS	HS	NPB, No Welt
102	3.00	RT	NPB, SDL	Browntone	Cigar	HS	HS	HS	SH, NPB, RW
101	2.50	RT	NPB, SDL	Natural	Cigar	HS	HS	HS	SH, NPB, RW
100	2.00	RT	NPB, SDL	Natural	Cigar	HS	HS	HS	SH, NPB, RW
171	7.50	RT	NPB, SDL	Flamed	Fat Half Wells	HS	HS	HS	SH, NPB, RW

Weber Lifelike Fly Company

Weber was located in Stevens Point, Wisconsin. The company was founded by O.L. Weber in 1919. Shortly after its inception, the Weber Co. acquired the C. J. Frost Fishing Tackle Company which had been in business since 1896. They also absorbed the Kinnickinnic Tackle Company and the Ross Artificial Fly Company of Portland, Oregon. Weber built a new plant in 1927 at Stevens Point to accomodate the new lines. By 1938, O.L. Weber's daughter, Helen Weber, was Superintendant of the factory and eventually took over the business upon the death of her father. Miss Weber had been an ardent fly angler since she caught her first bluegill on a fly rod as a small girl.

Weber used two slogans that were noteworthy. The best known is "IF WEBER MAKES IT, A FISH TAKES IT". This is notable only because it is clever (It has a beat and you can dance to it, as they say.) The truly notable slogan was printed at the bottom of *every* page of the Weber catalogs beginning in 1930. That historic slogan, as valuable today as it was then, was: "BUY AMERICAN, OR AMERICAN LABOR CANNOT BUY."

With all of their acquisitions, Weber was unable to buy a good rodmaking firm. From 1920 to 1929, Weber bought rods from Edwards. Beginning in 1930, Weber bought its rods from Heddon, and continued to do so throughout its history. The Heddon-made Weber rods are the most often seen, and are better rods than those that were made by Edwards.

As did most companies, Weber used a *famous angler* to help promote their rods and reels. Weber's expert was Bob Becker, a nationally known angling authority, writer and sportsman. He was the author of *MEMO: Go Fishing*" and *The Fisherman's Guide*. Rather than having his name on the rods, Weber used Becker in a novel way: He assembled (from the Weber line of tackle) a complete outfit of perfect balance. He selected a rod, reel and line of the proper size. The *matched unit* came with a tag bearing the likeness of Bob Becker, and was personally autographed to certify its perfect match. Weber rods were marked with two decals. One decal read *Weber of Stevens Point*, the other was the name of the rod as shown on the flat map for the Weberkraft model. The Water Witch label was used on that model only. Water-Witch was a trade name of the C. J. Frost Company. All Weber rods were available in lengths from 8' to 9½', and all were three-piece rods. Heavy action rods for Salmon or Steelhead were available by special order only.

Weber
of Stevens Point
"The Fly-Tackle Capital of America"

Weber
of Stevens Point

WEBERKRAFT

WATER
CFJ
WITCH

Weber Rods — Hardware & Wrap Specs — 1937

Model	Reelseat	Grip	Stripper	Snake	TT	Ferrules
No. 3500 Henshall MasterKraft	Heddon SDL	Rev. Half Wells	PS	PS	PS	GM, NS, HW, S
No. 2500 Henshall MasterKraft	Heddon SDL	Rev. Half Wells	PS	PS	PS	GM, NS, HW, S
No. 1500 Monogram	Heddon SDL	Rev. Half Wells	PS	PS	PS	NS, RW, w/o
No. 1200 WeberKraft	NPB, SB	Half Wells	PS	PS	PS	NS, RW, w/o
No. 1000 Water-Witch	NPB, SB	Rev Half Wells	PS	PS	PS	NS, RW, w/o

Model	$	Main Color	Tipping	Notes
Henshall MasterKraft	35.00	Brown #27	Purple #16	w/o WC, Open HK
Henshall MasterKraft	30.00	Brown #27	Dk. Brown #30	w/o WC, Open HK
Monogram	20.00	B&W Jasper	None	Exposed WC, Open HK
WeberKraft	12.00	Red #10	None	Exposed WC, Open HK
Water-Witch	10.00	B&O Jasper	None	Exposed WC, Open HK

Winchester Rods

Winchester Repeating Arms Company of New Haven, Conn. was an old and highly respected manufacturer and seller of sporting goods of all sorts. The products they sold were usually marked with the famous Winchester logo, though many were not actually made in their own factories.

In the period immediately following WW I, Edwards was approached by the Winchester Company about making rods to sell under the Winchester trademark. In 1919, the sale of E. W. Edwards Rod Co. was finalized, and as a condition of the sale Edwards went to New Haven, Conn. to supervise the manufacture of the Winchester Rods for five years. Eustis Edwards had been one of H. L. Leonard's proteges in New York, and had also been one of the principals in the Thomas, Edwards & Payne Co. which manufactured the Kosmic rods. As such, his name was associated with high-quality tackle.

It was perhaps the Edwards reputation as much as his rodmaking skills that made him the choice of Winchester to build the rods that would carry their trademark. This does indeed seem to be the case. From the very beginning, Winchester boasted that *These rods are made under the direct supervision of Mr. E. W. Edwards.* The highest grade rods (those selling for $50) were advertised as *Hand-made throughout by Mr. E. W.Edwards. These rods are perfect in every detail.* Edwards may not have had as much control over the rodbuilding operation as the catalogs lead us to believe. He became uncomfortable with the lower quality of the mass-produced rods, and when his contractual obligation was completed in 1924, he left Winchester and went into business on his own.

Edwards / Winchester Rods of higher grades were very good rods; they used high quality NS ferrules and hardware, and had good taper designs. The higher grades were the equal of any rods in production, and were better than most. The cane and finishing work was absolutely first-class. The middle grade rods were strictly average in all respects.

But, the lower grades had few redeeming qualities. The hardware used on the lowest grades was NPB, as were the ferrules. The cane work was typical of lower grade production rods, with random node spacing and little care in matching the cane for color and quality. The end result was a cheap rod that equalled the lower grades of Montague and Horrocks-Ibbotson. It is no wonder that Eustis Edwards soon tired of this type of business!

Markings on Edwards/Winchester Rods are usually only the model number, a four digit number that begins with the numeral "6", and corresponds to those shown on the chart. This number is usually stamped into the reelseat hardware, but may appear written on the rodshaft. Note that the model number designates the grade, length and type of action. Unless you have the chart as a reference, the model numbers are very confusing.

After Edwards left Winchester, the company continued to produce rods for several years without benefit of Edwards' expertise. The rods certainly did not get any better during this period. Without being able to use the Edwards name, Winchester needed a *name* to help sell its rods. In the late 1920's and very early 1930's,

Courtney Ryley Cooper, famous angling author, was the Winchester *poster child*. Full page, full color advertising was used in sporting magazines to announce that *Courtney Ryley Cooper uses Winchester Tackle.*

By 1935, Winchester had sold its rodmaking division, both bamboo and steel, to Horrocks-Ibbotson of Utica, N.Y. H-I / Winchester rods were of average or lower quality. The markings on these rods are usually simple. The Winchester name and model name are written on the rodshaft, with white ink, with the letters reading toward the grip. The rods themselves are standard H-I models which were re-named to help distinguish them as Winchesters. It does not seem that Winchester sold these rods. If that is true, Horrocks-Ibbotson made use of the Winchester reputation much as Winchester had exploited the Edwards name in the early 1920's. Wraps for H-I / Winchester rods varied dramatically, and almost any H-I wrap pattern may be used as long as you pay attention to the relative grade of the rod.

Winchester by Edwards — Hardware & Wrap Specs — 1920's

Rod #	$	Length	Action	Reelseat	Grip	Main Wrap	Trim Wrap
6186	50.00	10'	Dry Fly	NS, SB	Straight	Pongee #27	Purple #16
6181		9½'					
6176		9'					
6171		8½'					
6166		8					
6185	50.00	10'	Trout	NS, SB	Straight	Pongee #27	Purple #16
6180		9½'					
6175		9'					
6170		8½'					
6165		8'					
6161	35.00	9½'	Dry Fly	NS, SB	Straight	Pongee #27	None
6156		9'				(semi-transparent)	
6151		8½'					
6160	35.00	9½'	Trout	NS, SB	Straight	Pongee #27	None
6155		9'				(semi-transparent)	
6150		8½'					
6146	30.00	9½'	Dry Fly	NS, SB	Fat Cigar	Tan #27	None
6141		9'					
6136		8½'					

Rod #	$	Length	Action	Reelseat	Grip	Main Wrap	Trim Wrap
6145	30.00	9½'	Trout	NS, SB	Fat Cigar	Tan #27	None
6140		9'					
6135		8½'					
6131	25.00	9½'	Dry Fly	NS, SB	Fat Cigar	Green #24	None
6126		9'					
6121		8½'					
6116		8'					
6130	25.00	9½'	Trout	NS, SB	Fat Cigar	Green #24	None
6125		9'					
6120		8½'					
6115		8'					
6111	20.00	10'	Dry Fly	NS, SB	Full Wells	Black #1	Green #19
6106		9½'		(Frosted Finish)			
6101		9'					
6096		8½'					
6110	20.00	10'	Trout	NS, SB	Full Wells	Black #1	Green #19
6105		9½'		(Frosted Finish)			
6100		9'					
6095		8½'					

Winchester by Edwards **Hardware & Wrap Specs** **1920's**

Rod #	$	Length	Action	Reelseat	Grip	Main Wrap	Trim Wrap
6191	16.00	9½'	Dry Fly	NS, SB	Full Wells	Gray #31	Red #10
6186		9'					
6181		8½'					
6090	16.00	9½'	Trout	NS, SB	Full Wells	Gray #31	Red #10
6085		9'					
680		8½'					
6075	12.50	9½'	Trout	NS, SB	Slender Cigar	Black #1	None
6070		9'					
6065		8½'					
6060	10.00	10'	Bass	NPB, SB	Slender Cigar	Red #10	None
6055		9½					
6050		9'					
6045		9½'					
6040	8.00	9½'	Trout	NPB, SB	Slender Cigar	Purple #16	None
6035		9'					
6030		8½'					

Rod Hardware
A Comparative Guide

This chapter presents many different styles of hardware as used by most makers throughout the years. If you have ever had to try to describe hardware without being able to show it to someone, you will appreciate this chapter because now you can phone a buddy and say "My rod has guides like #xxx on page x". Although it seems repetitive, the author has elected to show different sizes of different styles so that if you are seeking a guide, ferrule or tip-top of a certain size, you can match up your original with the proper size from the book when trying to locate a duplicate or replacement.

Every effort has been made to show as many styles as possible, and many of the most commonly seen components are pictured. Note that winding checks are also called *tapers* in the tackle trade. Many styles and variations are seen, but pictured here are those most commonly seen. The hardware that appears in this chapter was readily available to all manufacurers – even amateurs. As a result, they are not usually reliable for determining the maker of a rod. You will note that some of the winding checks are the same as used on most Montague rods of the 1920's and 1930's, but that does not mean that every rod that uses the check is a Montague. Production and Trade Rod makers tended to use whatever was available at the best price.

I suppose that a good comparison today is athletic shoes; you'll see the premium brands in fancy mall stores selling for over $100, but cheap *knockoffs* are available in discount stores at much lower prices. The athletic shoe business is probably the only facet of business today that can match the intense competetive nature of the fishing tackle trade from 1890 to 1930. The final point here is that the cheap discount store shoes may look like the premium brands, but they just aren't of the same quality level. Likewise, it was beneficial to the makers of the cheapest rods to look like a better brand-name rod.

Butt Caps and Tapers.

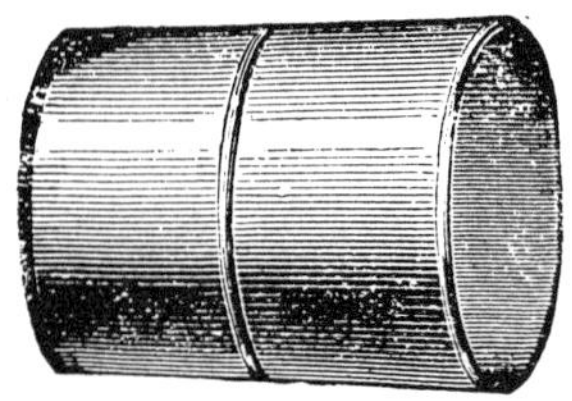

Solid End.

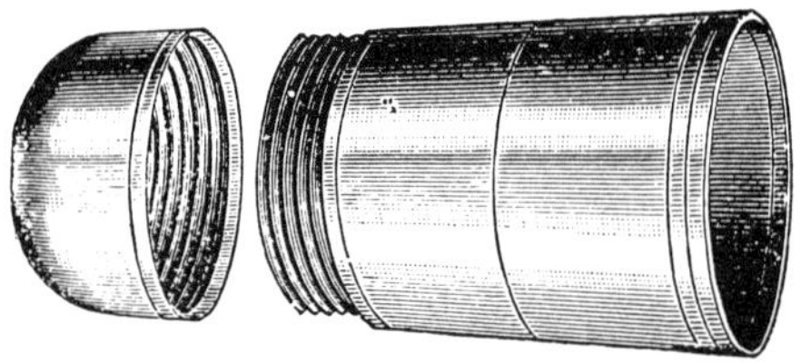

Screw-off.

Butt Caps.

Inch	¾	⅞	1	1 1/16	1⅛	1 3/16	1¼	
Solid End, Nickel Plated	$ .06	.07	.08	.09	.10	.11	.12	each
Solid End, German Silver	.12	.14	.16	.18	.25	...	...	"
Nickel Plated Screw-off	...	.11	.12	.14	.16	...	...	"

Tapers for Fly Rods.

Style of Nos. 4/0 to 3.

Nos.	4/0	3/0	2/0	1/0	1
Small End	12/32	14/32	15/32	16/32	15/32
Large End	18/32	18/32	18/32	18/32	19/32

Nos.	2	3	4	5
Small End	33/64	37/64	14/32	15/32
Large End	21/32	23/32	20/32	21/32

All sizes.

Nickel Plated .... $ .08 each

German Silver .... .15 "

Style of Nos. 4 and 5.

Tapers for Bait Rods.

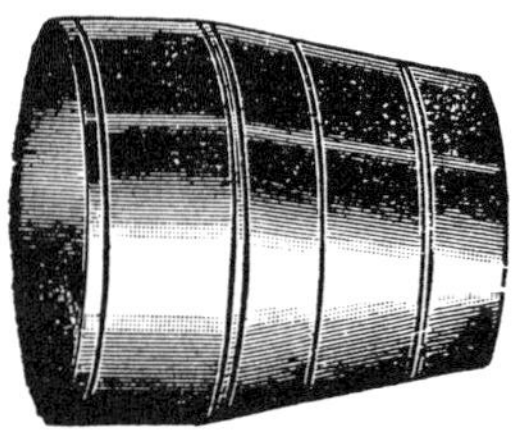

To fit Reel Seat	¾	⅞	1	1⅛	
Nickel Plated	$ .15	.15	.15	.15	each
German Silver	.20	.20	.20	.20	"

Winding Checks.

All sizes.

Diameter	13/16	15/16	1	1⅛	1¼

Nickel Plated .... $ .15 each

German Silver .... .20 "

REEL BANDS

No. 337

No. 334

No. 339

No. 333

No. 337	Plain or knurled German Silver reel band, ⅝ in., ¾ in., ⅞ in., 1 in. Each	$0.35
No. 437	Plain or knurled, Nickel Plate reel band, ⅝ in., ¾ in., ⅞ in., 1 in. Each	.10
No. 334	Roll Welt, German Silver reel band, ⅝ in., ¾ in., ⅞ in., 1 in. Each	.50
No. 434	Roll Welt, Nickel Plate reel band, ⅝ in., ¾ in., ⅞ in., 1 in. Each	.35
No. 436	Roll Welt Bronze finish reel band, ⅝ in., ¾ in., ⅞ in., 1 in. Each	.40
No. 335	Hand Welt, German Silver reel band, ¾ in., ⅞ in., 1 in., 1⅛ in. Each	.65
No. 435	Hand Welt, Nickel Plate reel band, ¾ in., ⅞ in., 1 in., 1⅛ in. Each	.50
No. 338	Reel Band with Hook, German Silver, ¾ in. only. Each	.55
No. 438	Reel Band with detachable Hook, Nickel, ¾ in. only. Each	.26
No. 339	Hooded Reel band (straight hoods), German Silver, ¾ in., ⅞ in., 1 in. Each	.50
No. 439	Hooded Reel band (straight hoods), Nickel Plated, ¾ in., ⅞ in., 1 in. Each	.30
No. 433	Hooded Reel band, Bronze finish, ¾ in., ⅞ in., 1 in. Each	.35
No. 332	"Quick Detachable" Finger Hook, German Silver, ¾ in. only. Each	.50
No. 333	Reel Clamp, German Silver (locking band), ¾ in., ⅞ in., 1 in., 1⅛ in. Each	.25

No. 340

No. 341

No. 347

No. 343

No. 344

No. 342

No. 349

BUTT CAPS, SOLID ENDS

Sizes		⅝	¾	⅞	1	1⅛
No. 340	German Silver. Each	....	$0.27	$0.34	$0.41	$0.54
No. 440	Nickel Plate. Each	....	.10	.12	.14	.17
No. 428	Bronze Finish. Each	....	.15	.16	.17	.22
No. 341	German Silver, Hooded. Each	$0.45	.45	.54	.65	....
No. 441	Nickel Plated, Hooded. Each	.22	.22	.30	.36	....
No. 429	Bronze Finish, Hooded. Each	.26	.26	.34	....	....

FLY ROD TAPERS

Figures denote diameters of each end in 64ths of an inch.

No. 343	German Silver, sizes: 40-20, 38-26, 38-30, 42-33, 46-37. Each	$0.27
No. 443	Nickel Plate, sizes: 40-20, 38-26, 42-33, 38-30, 46-37. Each	.14
No. 445	Bronze Finish, sizes: 40-20, 38-26, 38-30, 42-33, 46-37. Each	.13
No. 347	Nickel Silver, sizes: 40-22, 40-24, 40-26, 40-28, 40-30. Each	.11
No. 344	German Silver, sizes: 40-25, 44-33. Each	.27
No. 444	Nickel Plated, sizes: 40-25, 44-33. Each	.14

Ferrules.

Shouldered Ferrule.

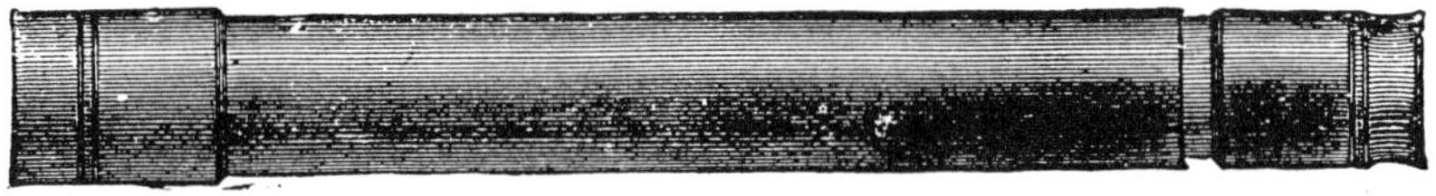

Welted Ferrule.

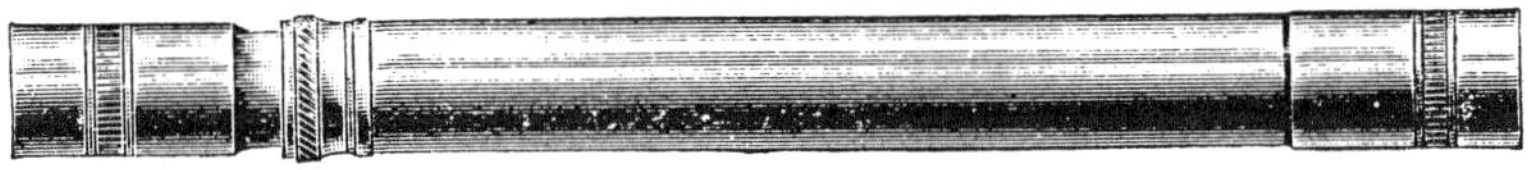

Serrated Ferrule.

Please order by number Nos.	Inside Diameter of Outside Ferrule	Nickel Plated, Shouldered Per pair	Welted Nickel Plated, Shouldered Per pair	Welted German Silver, Shouldered Per pair
3/0	9/64	....	....	$ .30
2/0	11/64	$ .08	$ .17	.30
0½	3/16	.08	.17	.30
1/0	7/32	.08	.17	.30
1	15/64	.08	.17	.30
1½	¼	.09	.18	.34
2	17/64	.09	.18	.34
2½	19/64	.09	.19	.36
3	11/32	.09	.19	.36
4	3/8	.10	.20	.38
5	27/64	.10	.20	.40
6	15/32	.11	.21	.42
7	½	.11	.21	.44
8	35/64	.12	.22	.46
9	19/32	.12	.22	.48
10	5/8	.13	.23	.50
11	43/64	.14	.24	.54
12	23/32	.15	.25	.58
13	¾	.16	.26	.62
14	51/64	.17	.27	.70
15	55/64	.19	.29	.74
16	7/8	.21	.31	...
17	61/64	.23	.33	...
18	63/64	.25	...	...
19	1 1/32	.27	...	...

Please Order by Number and Name.

Narrow Agate Guides.

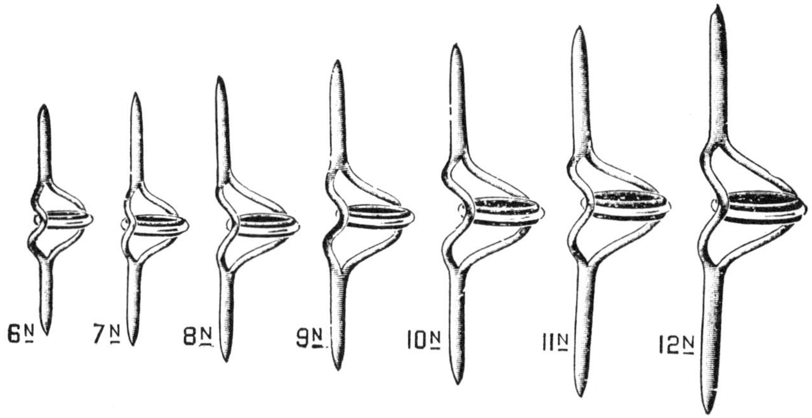

Very Light Weight for Fly Rods or Bait Casting Rods.

Nos. 6 N, 7 N, 8 N, 9 N, 10 N, 11 N, 12 N.

Each........................ $ .60   Per dozen.................. $6.00

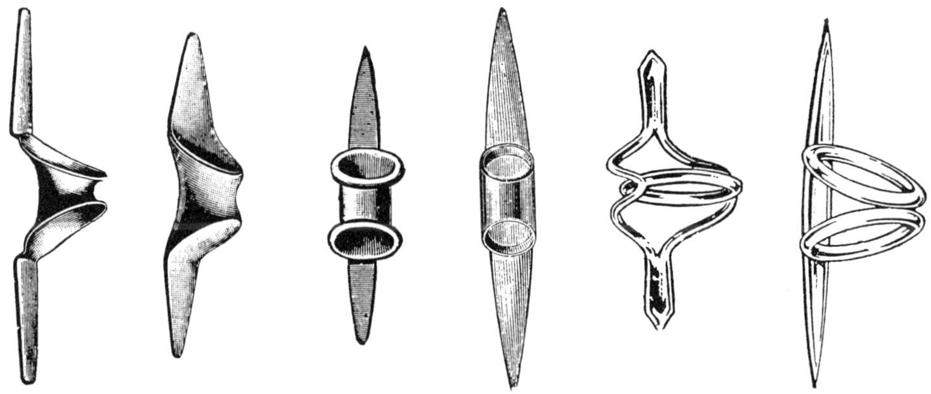

322   422   325   425   329   326

No. 322	Hand Made German Silver Bell Guide. Sizes, 4, 5, 6, 7, 8, 9, 10. Each.............	$0.08
No. 422	Stamped Nickeled Tie Guide. Sizes, 4, 5, 6, 7, 8, 9. Each..........................	.05
No. 325	Anti-Friction German Silver Trumpet Guide. Sizes, 4, 5, 6, 7, 8, 9. Each...........	.14
No. 425	Stamped Nickeled Tube Guide. Sizes, 4, 5, 6, 7, 8, 9. Each.........................	.03
No. 329	One Ring German Silver Casting Guide. Sizes, 6, 8, 12, 16, 20. Each.............	.12
No. 326	Two Ring German Silver Guides. Sizes, 4, 6, 8, 12, 16, 20. Each................	.03
No. 328	Three-Coil Spiral Guides, German Silver. Sizes, 3, 4, 5, 6, 7, 8. Each...........	.02

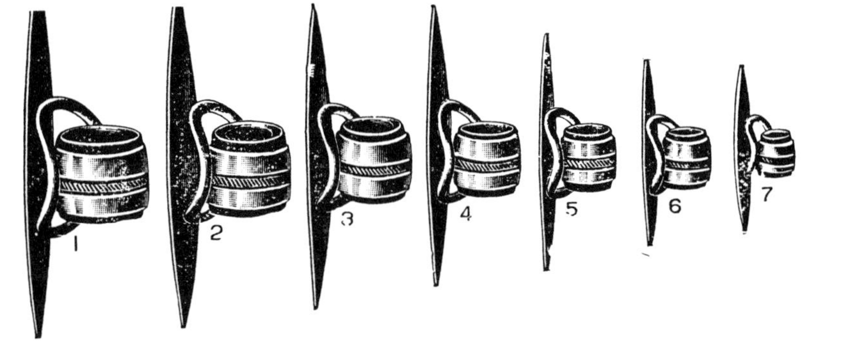

Flat Agate Guides.

Nos.	1	2	3	4	5	6	7
Each	$ .60	.60	.55	.50	.50	.50	.50

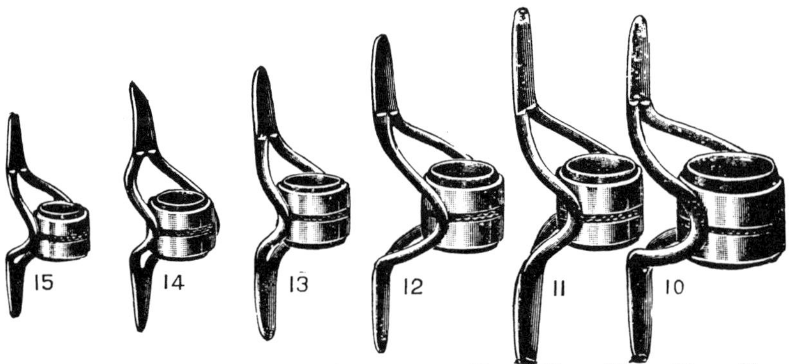

Wide Raised Agate Guides.

Nos.	10	11	12	13	14	15
Each	$ .95	.85	.80	.75	.70	.70

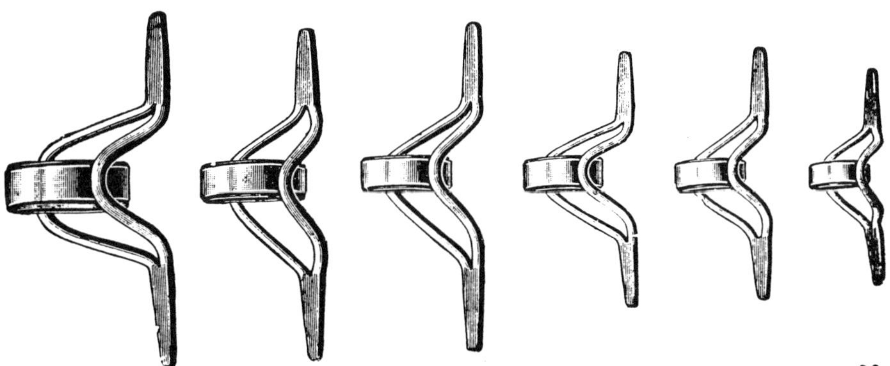

31 32 33 34 35 36

Medium Raised Agate Guides.

Each	$1.00	.85	.80	.75	.75	.75
Nos.	31	32	33	34	35	36

Hand-made German Silver Bell Guides.

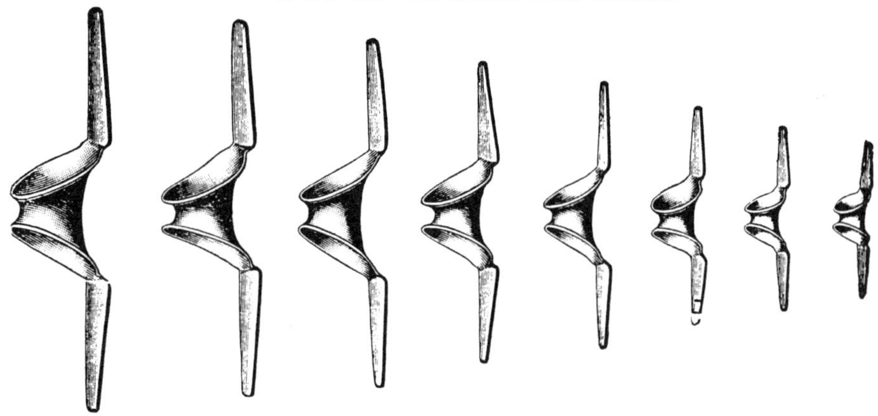

No. 1 2 3 4 — Per dozen........$1.50

5 6 7 8 — $1.20

German Silver Anti-Friction Trumpet Guides.

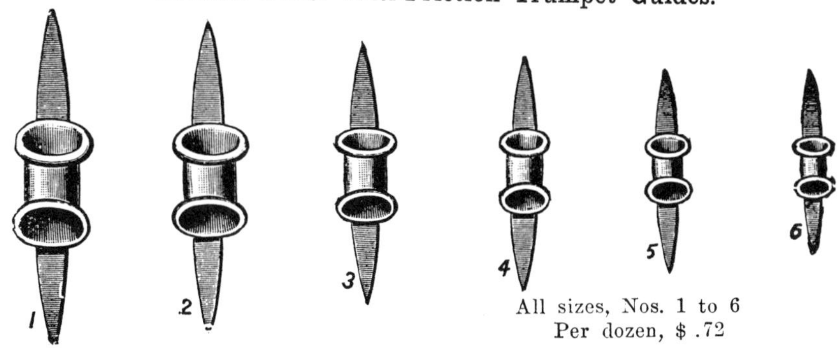

All sizes, Nos. 1 to 6
Per dozen, $ .72

Solderless Tie Guides.

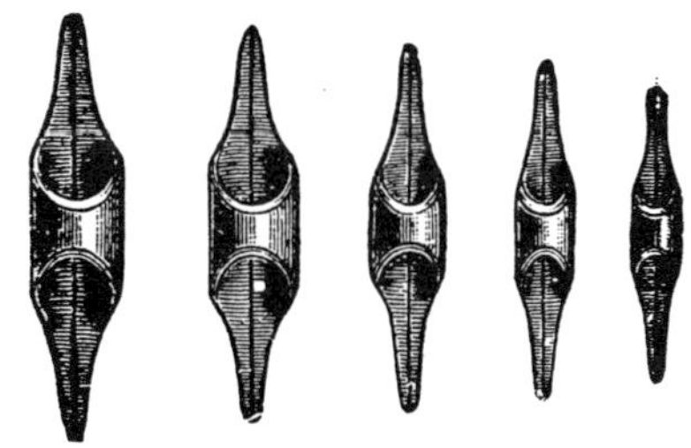

1 2 3 4 5

Brass, $ .18 doz. **Nickel plated,** $ .24 doz.

Three Coil Spiral Guides.

Nickel Plated......... $ .18 per dozen

German Silver Casting Tops and Guides.

(Numbers of tops indicate diameter of tube in 32nds of an inch.)

See illustration of sizes on page 11.

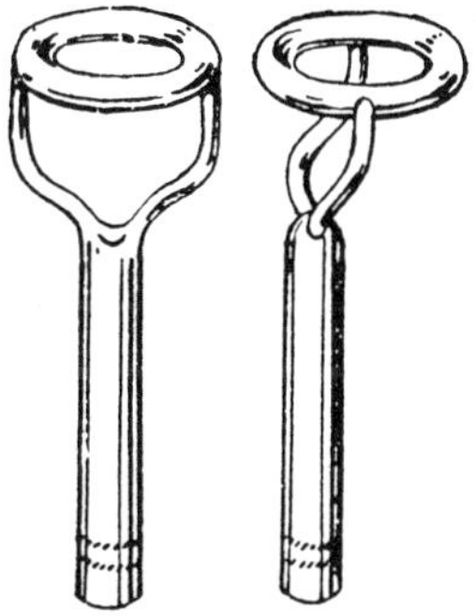

Offset Casting Top.

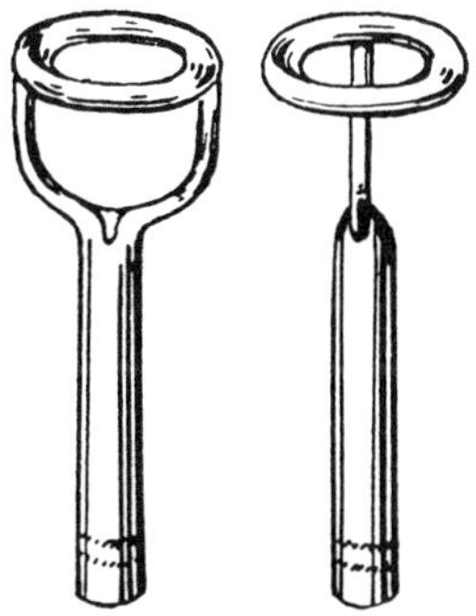

Casting Stirrup Top.

German Silver Casting Stirrup Tops, Nos. 4, 4½, 5, 6 Each, $ .12 Per doz., $1.20
German Silver Offset Casting Tops, Nos. 4, 4½, 5, 6 " .15 " 1.50

Two-Ring Casting Guides.

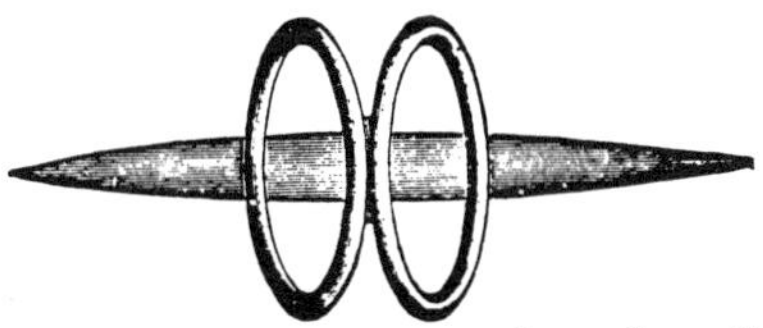

Nos.	1	2	3
Diameter of ring	⅜	½	⅝

All sizes. Each, $ .05. Per dozen, $ .50

Three-Coil Spiral Casting Guides.

Nos.	1	2	3	4
Diameter of ring	¼	5/16	⅜	7/16

All sizes. Each, $ .03. Per dozen, $ .30

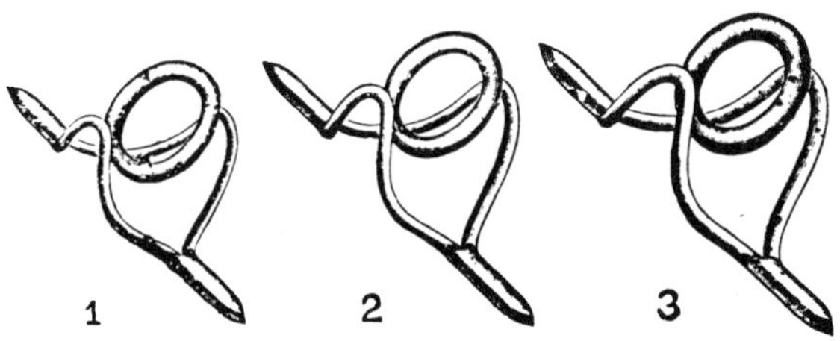

One Ring Casting Guides.

Nos.	0, 1, 2, 3
Each	$ .10
Per dozen	1.00

Illustration of Sizes of Tops.

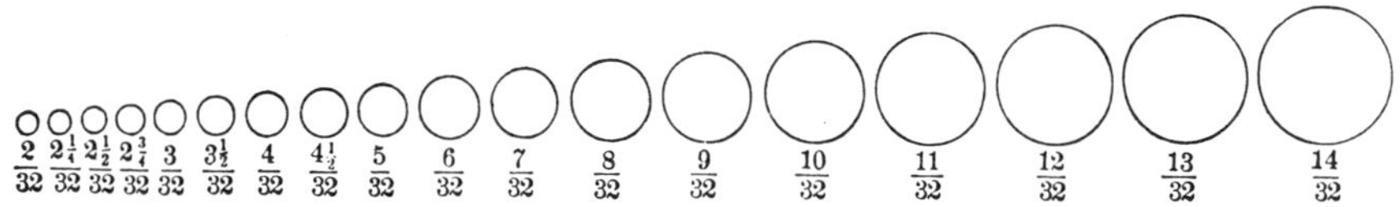

The numbers of Tops indicate the diameter of the tubes in 32nds of an inch.

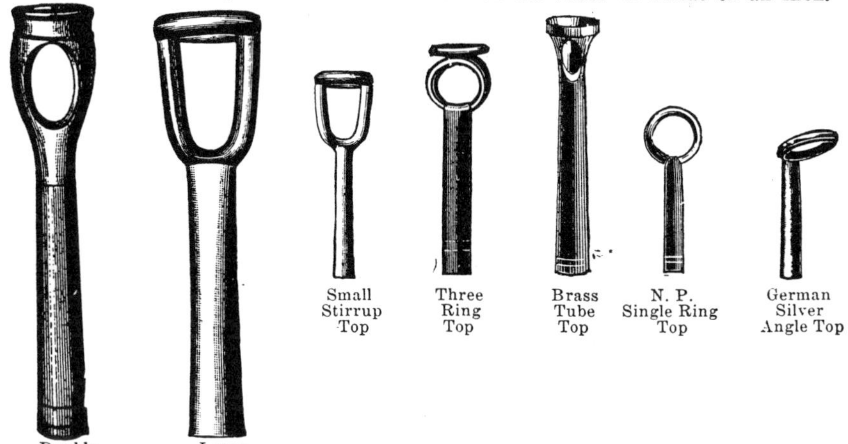

Double Hole Top — Large Stirrup Top — Small Stirrup Top — Three Ring Top — Brass Tube Top — N. P. Single Ring Top — German Silver Angle Top

Snake Guides.

German Silver, or Bronzed Steel.

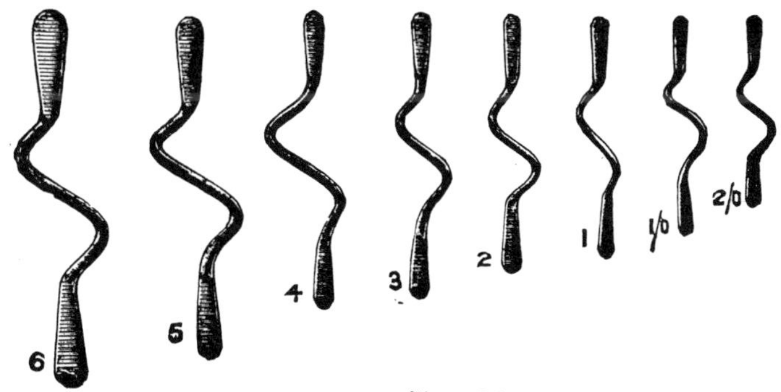

Sizes 3/0 to 6
Per dozen, $ .18. Per gross, $1.80

AGATE AND IMITATION AGATE TOPS AND GUIDES

All are Nickel Silver mounted and soldered, securely fastened. The Agates are cut from genuine stones, accurately fitted and well polished. The Imitation Agates, called Agatene, closely imitate the genuine stones, mounted the same, and will give excellent service. All tops and guides are made of 18% Silvore wire, silver (hard) soldered throughout. Smooth joints, polished and then nickel plated. All sizes are in 32nds of an inch. Guide sizes being inside diameter of ring or stones. Top sizes are inside diameter of ends of tubes.

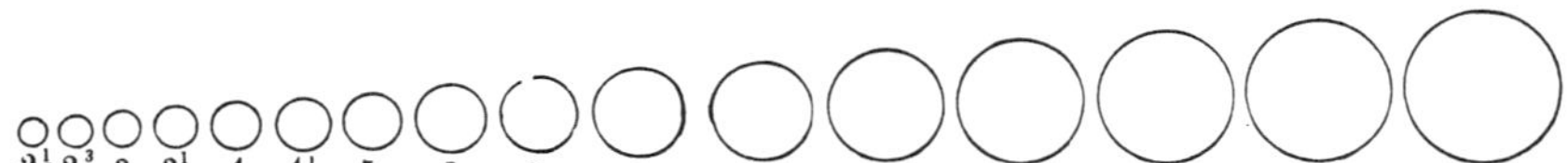

No. 374

No. 375

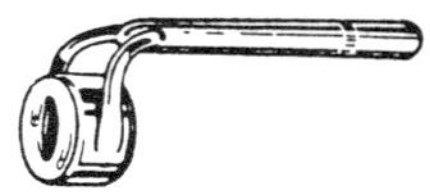

No. 372

SEAMLESS TUBE TOPS FOR FLY OR BAIT ROD

Sizes		2	2½	3	3½	4	4½	5	5½	6
No. 374	Fly Top, Agate	$0.36	$0.36	$0.36	$0.36	$0.36	...	...	...	...
No. 474	Fly Top, Agatene	.20	.20	.20	.20	.20	...	...	...	...
No. 375	Angle Top, Agate	...	...	.42	.42	.42	$0.42	$0.42	...	...
No. 475	Angle Top, Agatene	...	...	.26	.26	.26	.26	.26	$0.26	...
No. 372	Offset Top, Agate	...	...	.42	.42	.42	...	.42	...	$0.42
No. 472	Offset Top, Agatene	...	...	.22	.22	.22	.22	.22	.22	.22
No. 379	Guarded Top Agate	...	...	...	...	.52	.52	.52	...	.52
No. 479	Guarded Top, Agatene	...	...	...	.32	.32	.32	.32	...	...
No. 370	Stirrup Top, Agate	...	...	.44	.44	.44	.44	.44	...	.44
No. 470	Stirrup Top, Agatene	...	...	.24	.24	.24	.24	.24	...	.24

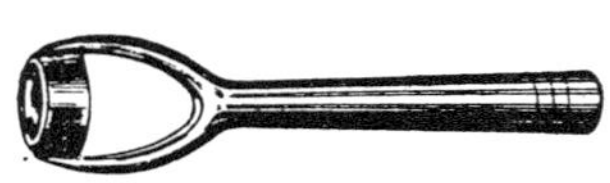

No. 370

No. 379

TAPERED TUBE TOPS FOR BAIT CASTING OR SALT WATER

Sizes		7	8	9	10	11	12
No. 376	Heavy Offset Top, Agate	$0.42	$0.42	$0.42	$0.56	$0.56	$0.56
No. 371	Stirrup Top, Agate, heavy construction	.56	.56	.56	.66	.66	.66
No. 471	Stirrup Top, Agatene, heavy construction	.38	.38	.38	.38	.38	.38
No. 378	Revolving Surf Top, Guarded, Agate	...	.90	.90	.90	.90	.90
No. 478	Revolving Surf Top, Guarded, Agatene	.88	.88	.88	.88	.88	.88
No. 377	Stirrup Top, Agate, extra heavy. Sizes 13, 14, 15, 16. Each						.66
No. 477	Stirrup Top, Agatene, extra heavy. Sizes 13, 14, 15, 16. Each						.40

No. 356

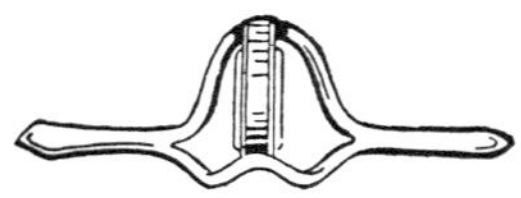

No. 350

No. 353

Suitable for Fly, Bait and Bait Casting Rods.

Sizes		5	6	7	8	10
No. 356	Agate, grooved, wire mounted guide	$0.39	$0.39	$0.42	$0.42	$0.42
No. 350	Agate, narrow banded	.44	.44	.46	.46	.50
No. 450	Agatene, narrow banded	.18	.20	.20	.22	...

Suitable for Bait Casting or Salt Water Rods

Sizes		12	14	16	18	20
No. 353	Agate, wide banded guides	$0.56	$0.56	$0.64	$0.80	$0.84
No. 453	Agatene, wide banded guides	.30	.34	.34	.34	.38

Shipping weight any item on this page 2 oz.

Split Rings and Guides for Fly Rods.

Steel Split Rings.

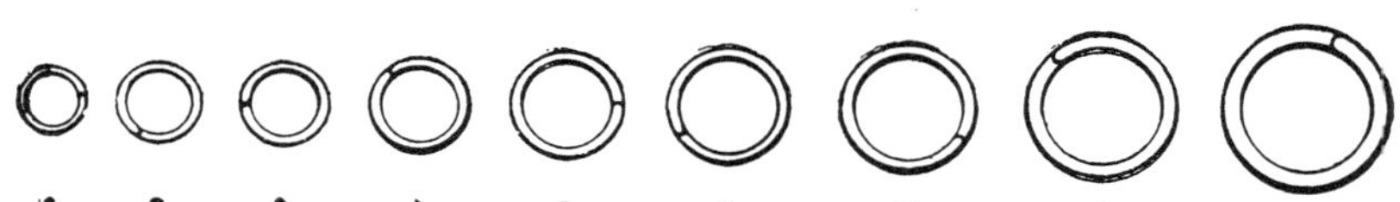

1 2 3 4 5 6 7 8 9

Nos.	1	2	3	4	5	6	7	8	9
Per dozen	$ .05	.05	.05	.05	.05	.06	.07	.08	.10

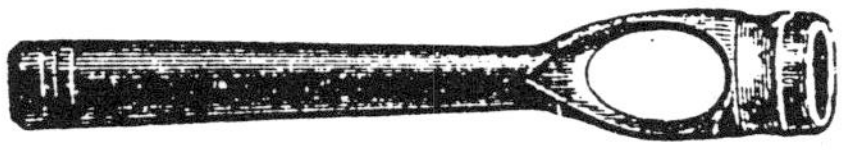

Style No. H, Nickel Plated

No. H—Sizes from 7 to 28. Each __20c

Style No. G

German Silver

No. G—Sizes 14 to 28. Each _____20c

Style No. D

No. D—Brass, 6 to 24. _10c

No. DN—Nickeled _____15c

SWIVEL AGATE TIP

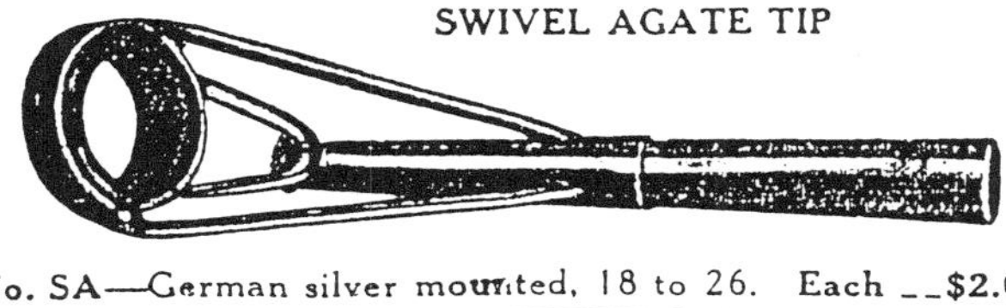

No. SA—German silver mounted, 18 to 26. Each __$2.00

Angle Agate No. L

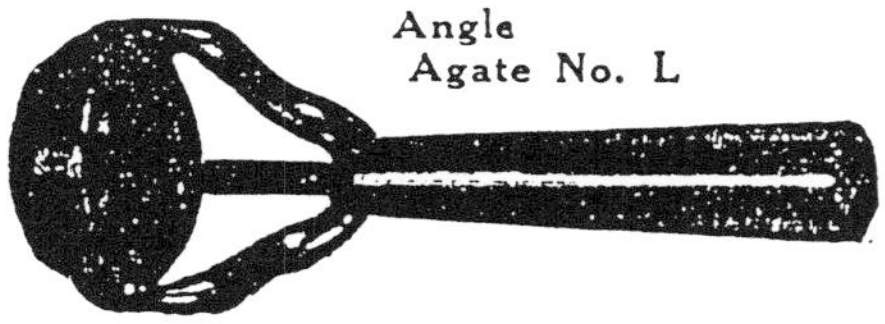

German Silver Mounted

No. L—Agate, sizes 18 to 24. Each..................75c

No. O—Agatine, sizes 18 to 24. Each................50c

No. GG—Plain G. S. offset. Each....................25c

Large Surf Casting—Style No. MS Agate German Silver Mounted

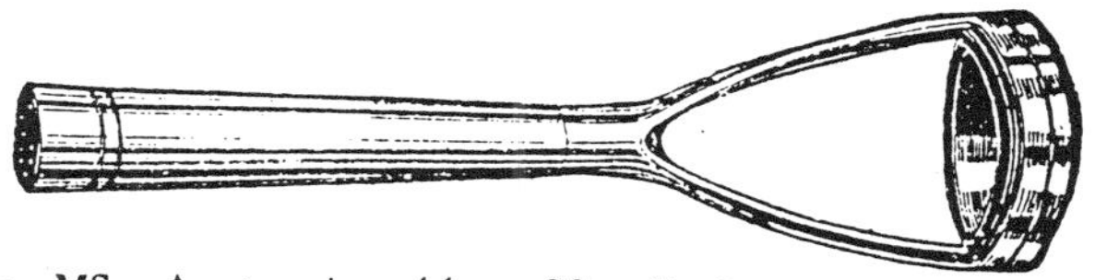

No. MS—Agate, sizes 16 to 28. Each __________$1.00

No. MC—Agatine, sizes 16 to 28. Each __________65c

Agate Stirrup Tip Style No. M .German Silver Mounted

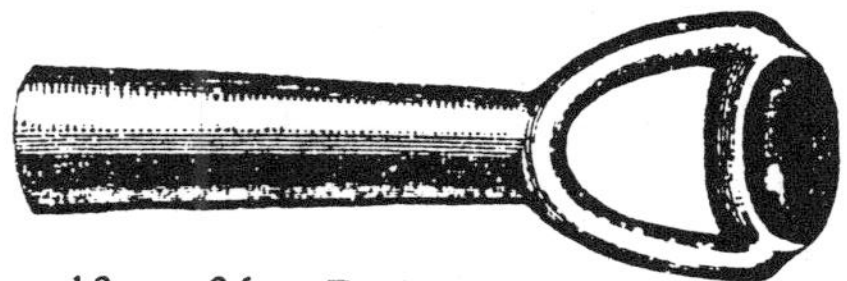

No. M—Agate, sizes 12 to 26. Each ____________75c

No. N—Agatine, sizes 14 to 28. Each ____________50c

Style No. I Casting

No. I—Sizes 7 to 12. Each ____10c

Style No. E

German Silver

No. E—6 to 14. Each ________10c

Carbon Steel Fly Tip Style No. A

No. A—Sizes 4 to 8. Each ____20c

No. AA—Chromium plated, Each, 25c

Agate Tip No. P German Silver

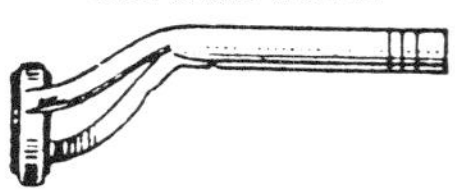

No. P—Agate, sizes 6 to 10 ___50c

No. W—Agatine, 6 to 12 _____25c

Agatine Tip No. Y

German Silver

No. Y—Sizes 7 to 12. Each ____25c

Style No. K German Silver

No. K—6 to 10. Each _________10c

Style No. C

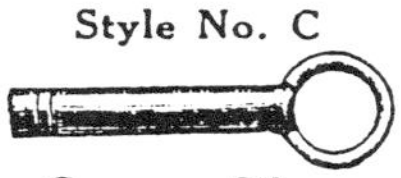

German Silver

No. C—4 to 8. Each __________5c

Agate Fly Tip—Style No. Q

No. Q—Sizes 3 to 7. Each ____40c

No. QQ—Agatine, 3 to 8. Each 20c

Agate Tip No. S German Silver

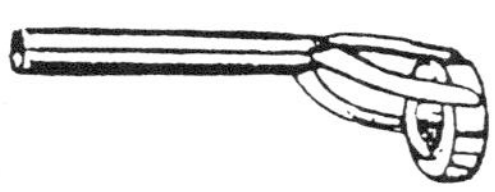

No. S—Agate, sizes 7 to 12 ___60c

No. SS—Agatine, 7 to 12 ______30c

Agate Tip for Steel Rod—No. XA

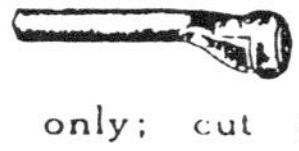

One size only; cut exact size; German silver.

No. XA—Agate. Each _______40c

No. X—Agatine. Each _______20c

Style No. F

German Silver

No. F—6 to 12. Each ________10c

Carbon Steel Casting Tip Style No. B

No. B—Sizes 6 to 12. Each ___25c.

No. BB—Chromium plated, Each, 35c

Agate Tip No. R German Silver

No. R—Agate, sizes 6 to 10. **Each 50c**

No. V—Agatine, 6 to 12. **Each 25c**

Agatine Tip No. NN

No. NN—Sizes 7 to 12 ________25c

Additional References

The Bamboo Rod – And How to Build It
By Claude M. Kreider
Centennial Publications, Grand Junction, CO; 1992

The Fine Bamboo Fly Rod
By Stuart Kirkfield.
Stackpole Books, Harrisburg, PA; 1986

The Wise Fishermen's Encyclopedia
Edited by A. J. McClane
Wm. H. Wise & Co., New York; 1957

McClane's Standard Fishing Encyclopedia
Edited by A. J. McClane
Holt, Rinehart & Winston, New York; 1965

Handcrafting Bamboo Fly Rods
By Wayne Cattanach
Self-published, 1993

A Master's Guide to Building a Bamboo Fly Rod
By Everett Garrison with Hoagy Carmichael
Meadow Run Press, New York: 1994

Idyll of the Split Bamboo
By George Holden Parker. Any Edition.

Fly Rods and Fly Tackle
By Henry P. Wells. Any Edition

About the Author

MICHAEL SINCLAIR is a historian who has lived in his adopted hometown of Denver for fourteen years. After graduating from Louisiana State University, he spent seventeen years in the restaurant business. For the past seven years he has remained *intentionally under-employed* to allow time for writing, research and as many hours on the stream as possible. When unable to go fishing, he spends his time restoring bamboo rods. Michael is a Mason, a member of Wheatridge Lodge No. 187, A.F. & A.M. and Peace Chapter No. 151, Order of the Eastern Star. Fortunately, his wife, Shorty, is not only his best friend and fishing partner, but also shares his enthusiasm for history, bamboo rods and is a member of Order of the Eastern Star.

Acknowledgments

My most heartfelt thanks: To Dick and Wendy Spurr for their constant patience, unflagging support and a few well-placed kicks. To my wife, Sharon Lee, for letting me. To Bob Corsetti for his encouragement and sharing of information. And most of all to the clients of the Cane Clinic, for allowing me to study, draw, and restore their rods.